Course **2**
Part **B**

Contemporary Mathematics in Context

A Unified Approach

CORE-PLUS MATHEMATICS PROJECT

Course
Part B **2**

Contemporary Mathematics in Context
A Unified Approach

Arthur F. Coxford
James T. Fey
Christian R. Hirsch
Harold L. Schoen
Gail Burrill
Eric W. Hart
Ann E. Watkins
with
Mary Jo Messenger
Beth E. Ritsema
Rebecca K. Walker

Glencoe
McGraw-Hill

New York, New York   Columbus, Ohio   Chicago, Illinois   Peoria, Illinois   Woodland Hills, California

Glencoe/McGraw-Hill

A Division of The **McGraw·Hill** *Companies*

 This project was supported, in part, by the National Science Foundation.
The opinions expressed are those of the authors and not necessarily those of the Foundation.

Send all inquiries to:

Glencoe/McGraw-Hill

8787 Orion Place

Columbus, OH 43240-4027

ISBN: 0-07-827541-5 (Part A)

ISBN: 0-07-827542-3 (Part B)

3 4 5 6 7 8 9 10 004/004 10 09 08 07 06 05 04 03

Contemporary Mathematics in Context

Course 2 Part B Student Edition

Core-Plus Mathematics Project Development Team

Project Directors

Christian R. Hirsch
Western Michigan University

Arthur F. Coxford
University of Michigan

James T. Fey
University of Maryland

Harold L. Schoen
University of Iowa

Senior Curriculum Developers

Gail Burrill
University of Wisconsin–Madison

Eric W. Hart
Western Michigan University

Ann E. Watkins
California State University, Northridge

Professional Development Coordinator

Beth E. Ritsema
Western Michigan University

Evaluation Coordinator

Steven W. Ziebarth
Western Michigan University

Advisory Board

Diane Briars
Pittsburgh Public Schools

Jeremy Kilpatrick
University of Georgia

Kenneth Ruthven
University of Cambridge

David A. Smith
Duke University

Edna Vasquez
Detroit Renaissance High School

Curriculum Development Consultants

Alverna Champion
Grand Valley State University

Cherie Cornick
Wayne County Alliance for Mathematics and Science

Edgar Edwards
(Formerly) Virginia State Department of Education

Richard Scheaffer
University of Florida

Martha Siegel
Towson University

Edward Silver
University of Michigan

Lee Stiff
North Carolina State University

Technical Coordinator

Wendy Weaver
Western Michigan University

Collaborating Teachers

Emma Ames
Oakland Mills High School, Maryland

Laurie Eyre
Maharishi School, Iowa

Joel Goodman
North Cedar Community High School, Iowa

Cheryl Bach Hedden
Sitka High School, Alaska

Michael J. Link
Central Academy, Iowa

Mary Jo Messenger
Howard County Public Schools, Maryland

Valerie Mills
Ann Arbor Public Schools, Michigan

Jacqueline Stewart
Okemos High School, Michigan

Michael Verkaik
Holland Christian High School, Michigan

Marcia Weinhold
Kalamazoo Area Mathematics and Science Center, Michigan

Graduate Assistants

Diane Bean
University of Iowa

Judy Flowers
University of Michigan

Gina Garza-Kling
Western Michigan University

Robin Marcus
University of Maryland

Chris Rasmussen
University of Maryland

Rebecca Walker
Western Michigan University

Production and Support Staff

James Laser
Michelle Magers
Cheryl Peters
Jennifer Rosenboom
Anna Seif
Kathryn Wright
Teresa Ziebarth
Western Michigan University

Software Developers

Jim Flanders
Colorado Springs, Colorado

Eric Kamischke
Interlochen, Michigan

Core-Plus Mathematics Project Field-Test Sites

Special thanks are extended to these teachers and their students who participated in the testing and evaluation of Course 2.

Ann Arbor Huron High School
Ann Arbor, Michigan
  Ginger Gajar
  Brenda Garr

Ann Arbor Pioneer High School
Ann Arbor, Michigan
  Jim Brink
  Tammy Schirmer

Arthur Hill High School
Saginaw, Michigan
  Virginia Abbott
  Felix Bosco
  David Kabobel

Battle Creek Central High School
Battle Creek, Michigan
  Teresa Ballard
  Steven Ohs

Bedford High School
Temperance, Michigan
  Ellen Bacon
  Linda Martin
  Lynn Parachek

Bloomfield Hills Andover High School
Bloomfield Hills, Michigan
  Jane Briskey
  Cathy King
  Ed Okuniewski
  Linda Robinson
  Roger Siwajek

Brookwood High School
Snellville, Georgia
  Ginny Hanley
  Linda Wyatt

Caledonia High School
Caledonia, Michigan
  Jenny Diekevers
  Kim Drefcenski
  Thomas Oster

Centaurus High School
Lafayette, Colorado
  Eilene Leach
  Gail Reichert

Clio High School
Clio, Michigan
  Bruce Hanson
  Lee Sheridan
  David Sherry

Davison High School
Davison, Michigan
  Evelyn Ailing
  Wayne Desjarlais
  Dan Tomczak
  Darlene Tomczak

Dexter High School
Dexter, Michigan
  Kris Chatas
  Widge Proctor

Ellet High School
Akron, Ohio
  Marcia Csipke
  Jim Fillmore
  Scott Slusser

Firestone High School
Akron, Ohio
  Barbara Adler
  Barbara Crucs
  Jennifer Walls

Flint Northern High School
Flint, Michigan
  Al Wojtowicz

Goodrich High School
Goodrich, Michigan
  Mike Coke
  John Doerr

Grand Blanc High School
Grand Blanc, Michigan
  Charles Carmody
  Linda Nielsen

Grass Lake Junior/Senior High School
Grass Lake, Michigan
  Larry Poertner

Gull Lake High School
Richland, Michigan
  Darlene Kohrman
  Dorothy Louden

Kalamazoo Central High School
Kalamazoo, Michigan
  Gloria Foster
  Bonnie Frye
  Amy Schwentor

Kelloggsville Public Schools
Wyoming, Michigan
  Jerry Czarnecki
  Steve Ramsey
  John Ritzler

Midland Valley High School
Langley, South Carolina
  Kim Huebner
  Janice Lee

Murray-Wright High School
Detroit, Michigan
  Jack Sada

North Lamar High School
Paris, Texas
  Tommy Eads
  Barbara Eatherly

Okemos High School
Okemos, Michigan
  Lisa Magee
  Jacqueline Stewart

Portage Northern High School
Portage, Michigan
  Pete Jarrad
  Scott Moore
  Jerry Swoboda

Prairie High School
Cedar Rapids, Iowa
  Dave LaGrange
  Judy Slezak

San Pasqual High School
Escondido, California
  Damon Blackman
  Gary Hanel
  Ron Peet
  Torril Purvis
  Becky Stephens

Sitka High School
Sitka, Alaska
  Mikolas Bekeris
  Cheryl Bach Hedden
  Dan Langbauer
  Tom Smircich

Sturgis High School
Sturgis, Michigan
  Craig Evans
  Kathy Parkhurst
  Dale Rauh
  Jo Ann Roe
  Kathy Roy

Sweetwater High School
National City, California
  Bill Bokesch
  Joe Pistone

Tecumseh High School
Tecumseh, Michigan
  Jennifer Keffer
  Elizabeth Lentz
  Carl Novak
  Eric Roberts

Traverse City High School
Traverse City, Michigan
  Diana Lyon-Schumacher
  Ken May

Vallivue High School
Caldwell, Idaho
  Scott Coulter
  Kathy Harris

Ypsilanti High School
Ypsilanti, Michigan
  Valerie Mills
  Don Peurach
  Kristen Stewart

Overview of Course 2

Part A

Unit 1 ▶ Matrix Models

Matrix Models extends student ability to use matrices and matrix operations to represent and solve problems from a variety of real-world settings while connecting important mathematical ideas from several strands.

Topics include matrix models in such areas as inventory control, social relations, archaeology, recidivism, ecosystems, sports, tournament rankings, and Markov processes; matrix operations, including row sums, matrix addition, scalar multiplication, matrix multiplication, and matrix powers; properties of matrices; and matrix methods for solving systems of linear equations.

Unit 3 ▶ Patterns of Association

Patterns of Association develops student understanding of the strength of association between two variables, how to measure the degree of the relation, and how to use this measure as a tool to create and interpret prediction lines for paired data.

Topics include rank correlation, Pearson's correlation coefficient, cause and effect related to correlation, impact of outliers on correlation, least squares linear models, the relation of correlation to linear models, and variability in prediction.

Unit 2 ▶ Patterns of Location, Shape, and Size

Patterns of Location, Shape, and Size develops student understanding of coordinate methods for representing and analyzing relations among geometric shapes, and for describing geometric change.

Topics include modeling situations with coordinates, including computer-generated graphics; distance in the coordinate plane, midpoint of a segment, and slope; designing and programming algorithms; methods for solving systems of equations; coordinate and matrix models of isometric transformations (reflections, rotations, and translations) and of size transformations; and similarity.

Unit 4 ▶ Power Models

Power Models develops student ability to recognize data patterns that involve direct or inverse power variation, to construct and analyze those models and combinations such as quadratic models, and to apply those models to a variety of problems.

Topics include basic power models with rules of the form $y = ax^b$ and combinations of power models with other simple models; analysis of quadratic models and equations from tabular, graphic, and symbolic viewpoints; square root and cube root relations, and fractional power and radical expressions.

Overview of Course 2

Part B

Unit 5 ▶ Network Optimization

Network Optimization extends student ability to use vertex-edge graphs to represent and analyze real-world situations involving network optimization, including optimal spanning networks and shortest routes.

Topics include vertex-edge graph models, optimization, algorithmic problem solving, matrices, trees, minimal spanning trees, shortest paths, Hamiltonian circuits and paths, and Traveling Salesperson problems.

Lesson 1 *Finding the Best Networks*
Lesson 2 *Shortest Paths and Circuits*
Lesson 3 *Looking Back*

Unit 7 ▶ Patterns in Chance

Patterns in Chance develops student ability to understand and visualize situations involving chance by using simulation and mathematical analysis to construct probability distributions.

Topics include probability distributions and their graphs, Multiplication Rule for Independent Events, waiting-time (or geometric) distributions, expected value, rare events, summation notation, and an introduction to the binomial distribution.

Lesson 1 *Waiting Times*
Lesson 2 *The Multiplication Rule*
Lesson 3 *Probability Distributions*
Lesson 4 *Expected Value of a Probability Distribution*
Lesson 5 *Looking Back*

Unit 6 ▶ Geometric Form and Its Function

Geometric Form and Its Function develops student ability to model and analyze physical phenomena with triangles, quadrilaterals, and circles and to use these shapes to investigate trigonometric functions, angular velocity, and periodic change.

Topics include parallelogram linkages, pantographs, similarity, triangular linkages (with one side that can change length); sine, cosine, and tangent ratios, indirect measurement; angular velocity, transmission factor, linear velocity; periodic change, radian measure, period, amplitude, and graphs of trigonometric models of the form $y = A \sin (Bx)$ or $y = A \cos (Bx)$.

Lesson 1 *Flexible Quadrilaterals*
Lesson 2 *Triangles and Trigonometric Ratios*
Lesson 3 *The Power of the Circle*
Lesson 4 *Looking Back*

Capstone ▶ Forests, the Environment, and Mathematics

Forests, the Environment, and Mathematics is a thematic, two-week project-oriented activity that enables students to pull together and apply the important mathematical concepts and methods developed throughout the course.

Contents

Preface

The first three courses in the *Contemporary Mathematics in Context* series provide a common core of broadly useful mathematics for all students. They were developed to prepare students for success in college, in careers, and in daily life in contemporary society. Course 4 formalizes and extends the core program with a focus on the mathematics needed to be successful in college mathematics and statistics courses. The series builds upon the theme of *mathematics as sense-making*. Through investigations of real-life contexts, students develop a rich understanding of important mathematics that makes sense to them and which, in turn, enables them to make sense out of new situations and problems.

Each course in the *Contemporary Mathematics in Context* curriculum shares the following mathematical and instructional features.

■ *Unified Content* Each year the curriculum advances students' understanding of mathematics along interwoven strands of algebra and functions, statistics and probability, geometry and trigonometry, and discrete mathematics. These strands are unified by fundamental themes, by common topics, and by mathematical habits of mind or ways of thinking. Developing mathematics each year along multiple strands helps students develop diverse mathematical insights and nurtures their differing strengths and talents.

■ *Mathematical Modeling* The curriculum emphasizes mathematical modeling including the processes of data collection, representation, interpretation, prediction, and simulation. The modeling perspective permits students to experience mathematics as a means of making sense of data and problems that arise in diverse contexts within and across cultures.

■ *Access and Challenge* The curriculum is designed to make more mathematics accessible to more students while at the same time challenging the most able students. Differences in student performance and interest can be accommodated by the depth and level of abstraction to which core topics are pursued, by the nature and degree of difficulty of applications, and by providing opportunities for student choice on homework tasks and projects.

■ *Technology* Numerical, graphics, and programming/link capabilities such as those found on many graphing calculators are assumed and appropriately used throughout the curriculum. This use of technology permits the curriculum and instruction to emphasize multiple representations (verbal, numerical, graphical, and symbolic) and to focus on goals in which mathematical thinking and problem solving are central.

■ *Active Learning* Instructional materials promote active learning and teaching centered around collaborative small-group investigations of problem situations followed by teacher-led whole class summarizing activities that lead to analysis, abstraction, and futher application of underlying mathematical ideas. Students are actively engaged in exploring, conjecturing, verifying, generalizing, applying, proving, evaluating, and communicating mathematical ideas.

■ *Multi-dimensional Assessment* Comprehensive assessment of student understanding and progress through both curriculum-embedded assessment opportunities and supplementary assessment tasks supports instruction and enables monitoring and evaluation of each student's performance in terms of mathematical processes, content, and dispositions.

Unified Mathematics

Contemporary Mathematics in Context is a unified curriculum that replaces the traditional Algebra-Geometry-Advanced Algebra/Trigonometry-Precalculus sequence. Each course features important mathematics drawn from four strands.

The Algebra and Functions strand develops student ability to recognize, represent, and solve problems involving relations among quantitative variables. Central to the development is the use of functions as mathematical models. The key algebraic models in the curriculum are linear, exponential, power, polynomial, logarithmic, rational, and trigonometric functions. Modeling with systems of equations, both linear and nonlinear, is developed. Attention is also given to symbolic reasoning and manipulation.

The primary goal of the Geometry and Trigonometry strand is to develop visual thinking and ability to construct, reason with, interpret, and apply mathematical models of patterns in visual and physical contexts. The focus is on describing patterns with regard to shape, size, and location; representing patterns with drawings, coordinates, or vectors; predicting changes and invariants in shapes; and organizing geometric facts and relationships through deductive reasoning.

The primary role of the Statistics and Probability strand is to develop student ability to analyze data intelligently, to recognize and measure variation, and to understand the patterns that underlie probabilistic situations. The ultimate goal is for students to understand how inferences can be made about a population by looking at a sample from that population. Graphical methods of data analysis, simulations, sampling, and experience with the collection and interpretation of real data are featured.

The Discrete Mathematics strand develops student ability to model and solve problems involving enumeration, sequential change, decision-making in finite settings, and relationships among a finite number of elements. Topics include matrices, vertex-edge graphs, recursion, voting methods, and systematic counting methods (combinatorics). Key themes are discrete mathematical modeling, existence (Is there a solution?), optimization (What is the best solution?), and algorithmic problem-solving (Can you efficiently construct a solution?).

Each of these strands is developed within focused units connected by fundamental ideas such as symmetry, matrices, functions, and data analysis and curve-fitting. The strands also are connected across units by mathematical habits of mind such as visual thinking, recursive thinking, searching for and explaining patterns, making and checking conjectures, reasoning with multiple representations, inventing mathematics, and providing convincing arguments and proofs.

The strands are unified further by the fundamental themes of data, representation, shape, and change. Important mathematical ideas are frequently revisited through this attention to connections within and across strands, enabling students to develop a robust and connected understanding of mathematics.

Active Learning and Teaching

The manner in which students encounter mathematical ideas can contribute significantly to the quality of their learning and the depth of their understanding. *Contemporary Mathematics in Context* units are designed around multi-day lessons centered on big ideas. Lessons are organized around a four-phase cycle of classroom activities, described in the following paragraph—*Launch*, *Explore*, *Share and Summarize*, and *On Your Own*. This cycle is designed to engage students in investigating and making sense of problem situations, in constructing important mathematical concepts and methods, in generalizing and proving mathematical relationships, and in communicating both orally and in writing their thinking and the results of their efforts. Most classroom activities are designed to be completed by students working together collaboratively in groups of two to four students.

The launch phase promotes a teacher-led class discussion of a problem situation and of related questions to think about, setting the context for the student work to follow. In the second or explore phase, students investigate more focused problems and questions related to the launch situation. This investigative work is followed by a teacher-led class discussion in which students summarize mathematical ideas developed in their groups, providing an opportunity to construct a shared understanding of important concepts, methods, and approaches. Finally, students are given a task to complete on their own, assessing their initial understanding of the concepts and methods.

Each lesson also includes tasks to engage students in Modeling with, Organizing, Reflecting on, and Extending their mathematical understanding. These MORE tasks are central to the learning goals of each lesson and are intended primarily for individual work outside of class. Selection of tasks for use with a class should be based on student performance and the availability of time and technology. Students can exercise some choice of tasks to pursue, and at times they can be given the opportunity to pose their own problems and questions to investigate.

Multiple Approaches to Assessment

Assessing what students know and are able to do is an integral part of *Contemporary Mathematics in Context*, and there are opportunities for assessment in each phase of the instructional cycle. Initially, as students pursue the investigations that make up the curriculum, the teacher is able to informally assess student understanding of mathematical processes and content and their disposition toward mathematics. At the end of each investigation, the "Checkpoint" and accompanying class discussion provide an opportunity for the teacher to assess levels of understanding that various groups of students have reached as they share and summarize their findings. Finally, the "On Your Own" problems and the tasks in the MORE sets provide further opportunities to assess the level of understanding of each individual student. Quizzes, in-class exams, take-home assessment tasks, and extended projects are included in the teacher resource materials.

Acknowledgments

Development and evaluation of the student text materials, teacher materials, assessments, and calculator software for *Contemporary Mathematics in Context* was funded through a grant from the National Science Foundation to the Core-Plus Mathematics Project (CPMP). We are indebted to Midge Cozzens, Director of the NSF Division of Elementary, Secondary, and Informal Education, and our program officers James Sandefur, Eric Robinson, and John Bradley for their support, understanding, and input.

In addition to the NSF grant, a series of grants from the Dwight D. Eisenhower Higher Education Professional Development Program has helped to provide professional development support for Michigan teachers involved in the testing of each year of the curriculum.

Computing tools are fundamental to the use of *Contemporary Mathematics in Context*. Appreciation is expressed to Texas Instruments and, in particular, Dave Santucci for collaborating with us by providing classroom sets of graphing calculators to field-test schools.

As seen on page v, CPMP has been a collaborative effort that has drawn on the talents and energies of teams of mathematics educators at several institutions. This diversity of experiences and ideas has been a particular strength of the project. Special thanks is owed to the exceptionally capable support staff at these institutions, particularly at Western Michigan University.

From the outset, our work has been guided by the advice of an international advisory board consisting of Diane Briars (Pittsburgh Public Schools), Jeremy Kilpatrick (University of Georgia), Kenneth Ruthven (University of Cambridge), David A. Smith (Duke University), and Edna Vasquez (Detroit Renaissance High School). Preliminary versions of the curriculum materials also benefited from careful reviews by the following mathematicians and mathematics educators: Alverna Champion (Grand Valley State University), Cherie Cornick (Wayne County Alliance for Mathematics and Science), Edgar Edwards (formerly of the Virginia State Department of Education), Richard Scheaffer (University of Florida), Martha Siegel (Towson University), Edward Silver (University of Michigan), and Lee Stiff (North Carolina State University).

Our gratitude is expressed to the teachers and students in our 35 evaluation sites listed on pages vi and vii. Their experiences using pilot- and field-test versions of *Contemporary Mathematics in Context* provided constructive feedback and improvements. We learned a lot together about making mathematics meaningful and accessible to a wide range of students.

A very special thank you is extended to Barbara Janson for her interest and encouragement in publishing a core mathematical sciences curriculum that breaks new ground in terms of content, instructional practices, and student assessment. Finally, we want to acknowledge Eric Karnowski for his thoughtful and careful editorial work and express our appreciation to the staff of Glencoe/McGraw-Hill who contributed to the publication of this program.

To the Student

Contemporary Mathematics in Context, Course 2 builds on the mathematical concepts, methods, and habits of mind developed in Course 1. With this text, you will continue to learn mathematics by doing mathematics, not by memorizing "worked out" examples. You will investigate important mathematical ideas and ways of thinking as you try to understand and make sense of realistic situations. Because real-world situations and problems often involve data, shape, change, or chance, you will learn fundamental concepts and methods from several strands of mathematics. In particular, you will develop an understanding of broadly useful ideas from algebra and functions, from statistics and probability, from geometry and trigonometry, and from discrete mathematics. You also will see connections among these strands—how they weave together to form the fabric of mathematics.

Because real-world situations and problems are often open-ended, you will find that there may be more than one correct approach and more than one correct solution. Therefore, you will frequently be asked to explain your ideas. This text will provide you help and practice reasoning and in communicating clearly about mathematics.

Because solving real-world problems often involves teamwork, you often will work collaboratively with a partner or in small groups as you investigate realistic and interesting situations. You will find that two to four students working collaboratively on a problem can often accomplish more than any one of you would working individually. Because technology is commonly used in solving real-world problems, you will use a graphing calculator or computer as a tool to help you understand and make sense of situations and problems you encounter.

As in Course 1, you're going to learn a lot of useful mathematics—and it's going to make sense to you. You're going to learn a lot about working cooperatively and communicating with others as well. You're also going to learn how to use technological tools intelligently and effectively. Finally, you'll have plenty of opportunities to be creative and inventive. Enjoy!

Network Optimization

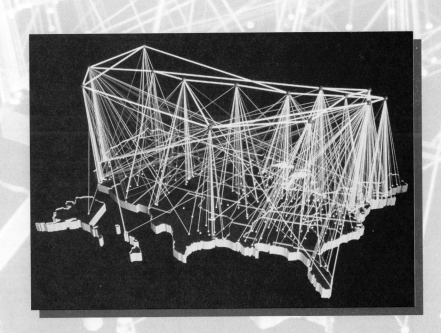

Finding the Best Networks

Everywhere you look, you'll see people trying to get the best in life, whether it's "going for the gold" in the Olympics, getting the best education, or even something as ordinary as finding the best route from home to school. Of course, what is best depends on the situation. In your previous work in mathematics, you may have investigated situations in which finding the "best" meant finding the best-fitting line for a scatterplot of data, the best plan for controlling population growth, the most stable geometric shape, or the fairest ranking of a tournament. In the context of vertex-edge graph models, you may have found the length of a longest path through a project digraph, the fewest number of colors needed to color the vertices of a graph, or the most efficient circuit through a graph using

all the edges. In this unit, you will continue to "find the best" in the context of graph models. In particular, you will study shortest paths and networks.

In this information age, it is important to find the best way to stay informed. You need to have the right information at the right time in order to make the best decisions and take the most effective action. One way to keep informed is through computer networks. Many places, including businesses, schools, and libraries, have computers linked together in networks so that information can be shared among many users. In fact, there is a common saying that "the network *is* the computer."

At a local high school, six computers in six different offices are to be networked. The school wants the best possible network.

(a) Why do you think it would be useful for the school to network the computers?

(b) What are some different ways to network the computers?

(c) Which way do you think is the best? Why?

INVESTIGATION 1 Optimizing a Computer Network

Suppose the school decides to link all the computers to each other without any kind of separate *junction box* or *server*. One way to get the "best" network is to use the least amount of wire to link all the computers. Since electronic signals move so quickly, the connection between two computers is virtually as efficient whether they are linked directly or indirectly through another computer. The person in charge of setting up the network wants to answer the following question:

What is the minimum amount of wire needed to connect all six computers so that every computer is linked directly or indirectly to every other computer?

Because of the locations of the offices and the computers, it is not possible to run wire directly between every pair of computers. The matrix below shows which computers can be linked directly, as well as how much wire is needed. The computers are represented by letters, and the distances are in meters.

	A	B	C	D	E	F
A	–	9	–	–	–	3
B	9	–	8	–	8	11
C	–	8	–	3	5	–
D	–	–	3	–	6	11
E	–	8	5	6	–	9
F	3	11	–	11	9	–

1. Examine the computer network matrix.

   a. What does the "5" in the *C* row mean?

   b. Why is the *A-B* entry the same as the *B-A* entry?

   c. Why isn't there a *D-D* entry? Why isn't there a *D-B* entry?

   d. Why does it make sense that the matrix is symmetric?

2. Represent the information in the matrix with a vertex-edge graph. Recall that when building a graph model you must specify what the vertices and edges represent.

3. Work with a partner to analyze your graph and the computer network problem.

   a. Compare your graph to your partner's graph. Resolve any differences.
      - What do the vertices represent?
      - What do the edges represent?

   b. What is the least amount of wire needed to connect the computers so that every computer is linked directly or indirectly to every other computer?

   c. Make a copy of the graph you agreed on in Part a, and then darken the edges of a shortest network.

   d. Write a description of a method for finding a shortest network.

   e. Compare your shortest network and the minimum amount of wire needed to what other students found. Discuss and resolve any differences.

4. In Activity 3, Part d, you and your partner wrote a description of a method for finding a shortest network.

   a. Exchange written descriptions with another pair of students. Make a new copy of the graph, and try to use the other pair's method to find a shortest network.

   b. Does each method work?

   c. Work together to refine the methods, and then write down a step-by-step procedure—an **algorithm**—for each method that works.

5. Think about the properties of the shortest wiring networks you have been investigating. State whether each of the following statements is *true* or *false*. In each case, give a reason justifying your answer. Compare your answers to those of other students and resolve any differences.

   a. There is only one correct answer possible for the minimum amount of wire needed to connect all six computers.

   b. There can be more than one shortest network for a given situation.

   c. There is more than one algorithm for finding a shortest network.

   d. A shortest network must be all in one piece; that is, the network must be **connected**.

   e. All vertices must be joined by the network.

   f. A shortest network cannot contain any circuits. (A **circuit** is a path that starts and ends at the same vertex and does not repeat any edges.)

6. A connected graph that has no circuits is called a **tree**.

a. Why does it make sense to call such a graph a tree?

b. A **minimal spanning tree** in a connected graph is a tree that has minimum total length and *spans* the graph—that is, it includes every vertex. Explain why the shortest networks you have found in the computer network graph are minimal spanning trees.

7. As you may have concluded in Activity 5, Part c, there are several possible algorithms for finding a minimal spanning tree in a connected graph. Study the algorithm below.

i. Draw all the vertices but no edges.

ii. Add the shortest edge that will not create a circuit. If there is more than one such edge, choose any one. The edge you add does not have to be connected to previously-added edges, and you may use more than one edge of the same length.

iii. Repeat Step ii until it is no longer possible to add an edge without creating a circuit.

a. Follow the steps of this algorithm to construct a minimal spanning tree for the computer network graph.

b. Explain why this algorithm could be called a *best-edge algorithm*.

c. Compare the minimal spanning tree you get using this best-edge algorithm to the one you found in Activity 3. How do the lengths of the minimal spanning trees compare?

8. Examine the graph below.

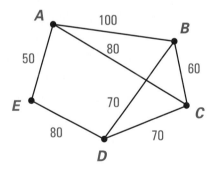

a. Use the best-edge algorithm to find a minimal spanning tree for this graph. Calculate its length.

b. Explain why the algorithm can produce different minimal spanning trees.

c. Find all possible minimal spanning trees for the graph. Compare their lengths.

d. How is the algorithm similar to or different from the algorithms you produced in Activity 4?

9. Students in one class claimed that the following algorithm will produce a minimal spanning tree in a given graph.

    i. Make a copy of the graph with the edges drawn lightly.

    ii. Choose a starting vertex.

    iii. From the vertex where you are, darken the shortest edge that will not create a circuit. (If there is more than one such edge, choose any one.) Then move to the end vertex of that edge.

    iv. Repeat Step iii until all vertices have been reached.

Complete Parts a–f to test this algorithm. First make four copies of the graph below.

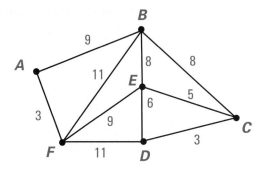

a. Apply the algorithm to the graph starting with vertex *E*. What is the total length of the network you get?

b. Explain why this algorithm could be called a *nearest-neighbor algorithm*.

c. Apply the algorithm starting with vertex *C*. Record the length of the resulting network.

d. Apply the algorithm starting with vertex *A*. What happens?

e. Now use the best-edge algorithm described in Activity 7 to find a minimal spanning tree for this graph.

f. Do you think the nearest-neighbor algorithm is a good algorithm for finding a minimal spanning tree? Write a brief justification of your answer.

10. How are the nearest-neighbor algorithm and the best-edge algorithm similar? How are they different?

11. Two important questions about any algorithm are "Does it always work?" and "Is it efficient?" You will continue to investigate these questions for different algorithms throughout this unit. For the best-edge algorithm described in Activity 7, mathematicians have proven that the answer to both questions is "yes." The best-edge algorithm will efficiently find a minimal spanning tree for any connected graph. What are your thoughts about these questions for the nearest-neighbor algorithm?

a. Does it always work?

b. Is it efficient?

The best-edge algorithm that you investigated in Activities 7 and 8 was first published by Joseph Kruskal, a mathematician at AT&T Bell Laboratories, and so it is also called *Kruskal's algorithm.* Kruskal discovered the algorithm while still a graduate student in the 1950s.

▶ On Your Own

A landscape architect has been contracted to design a sprinkler system for a large lawn. There will be six sprinkler heads that must be connected by a buried network of pipes to the main water source. The possible connections and distances in yards are shown in the diagram below. The main water source is represented by vertex *B*. What is the least amount of pipe needed to construct the sprinkler network? Draw a landscape plan showing the optimal sprinkler network.

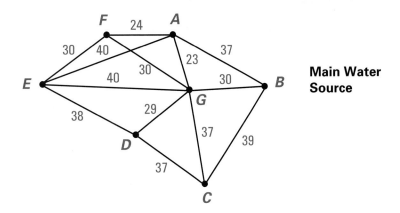

Main Water Source

INVESTIGATION 2 Optimizing a Road Network

You can use graph models to optimize many different kinds of networks. For example, consider the following road network. There are seven small towns in Johnson County that are connected to each other by gravel roads, as in the following diagram. (The diagram is not drawn to scale and the roads often are curvy.) The distances are given in miles. The county, which has a limited budget, wants to pave some of the roads so that people can get from every town to every other town on paved roads, either directly or indirectly, and yet the total number of miles paved is the minimum possible.

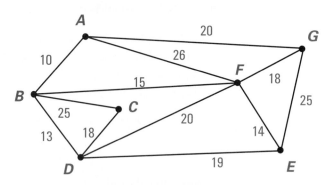

1. Find and draw a network of paved roads that will fulfill the county's requirements. Eliminate any unpaved roads from your drawing.

2. Construct a *distance matrix* for your paved-road network by completing a copy of the matrix below. The entries give the shortest distance between towns on the paved-road network that you found, not on the original gravel-road network shown above.

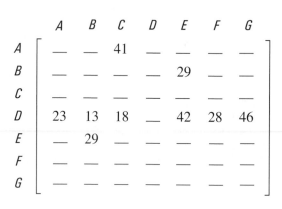

	A	B	C	D	E	F	G
A	—	—	41	—	—	—	—
B	—	—	—	—	29	—	—
C	—	—	—	—	—	—	—
D	23	13	18	—	42	28	46
E	—	29	—	—	—	—	—
F	—	—	—	—	—	—	—
G	—	—	—	—	—	—	—

a. Explain why the *A-C* entry is 41 and the *D-E* entry is 42.

b. Why is the *B-E* entry the same as the *E-B* entry?

c. Fill in all the entries of the matrix. Divide this job among your group.

d. Describe and explain any patterns you see in the matrix.

3. Use the distance matrix to further analyze the road network.

   a. Which two towns are farthest apart on the paved-road network?

   b. Compute the row sums of the distance matrix. What information do the row sums give about distances on the paved-road network?

   c. Which town seems to be most isolated on the paved-road network? Which town seems most centrally located? Explain how these questions can be answered by examining the distance matrix.

4. Which towns might be dissatisfied with this paved-road network? Why? What are some other considerations that might be taken into account when planning an optimal paved-road network?

Checkpoint

Suppose you find a minimal spanning tree for a graph that represents a road network and then construct the corresponding distance matrix.

a What information does the distance matrix for the minimal spanning tree give you?

b What is some useful information about the road network that you cannot get from the distance matrix?

c What information about towns and distances do the row sums give you?

Be prepared to share your ideas with the class.

On Your Own

In Investigation 1, you optimized a computer network by finding a network that used the least amount of wire. Another factor that you might want to optimize is cost. Consider the same six computers in the same six offices. The matrix below shows the *cost*, in dollars, to make each possible direct connection.

	A	B	C	D	E	F
A	–	39	–	–	–	33
B	39	–	38	–	39	41
C	–	38	–	44	41	–
D	–	–	44	–	36	37
E	–	39	41	36	–	36
F	33	41	–	37	36	–

a. Represent the information in the matrix with a vertex-edge graph.

b. Use the best-edge algorithm to find a network that connects all six computers, either directly or indirectly, for the least total cost.

c. Construct the "distance" matrix for your minimal spanning tree (similar to the one you constructed for Activity 2 of Investigation 2). Note that in this case, the entries of the "distance" matrix actually show the *cost* of the connection between each pair of vertices on the least-cost network.

d. Do you think that the row sums of the "distance" matrix provide any relevant information about the least-cost network? Explain.

MORE

Modeling • Organizing • Reflecting • Extending

Modeling

1. The graph to the right shows a road network connecting six towns (not shown to scale). The distances shown are in miles. The Highway Department wants to plow enough roads after a snowstorm so that people can travel from any town to any other town on plowed roads. However, because of the time and cost involved, officials want to plow as few miles of road as possible.

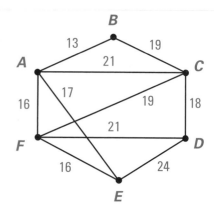

a. Find and draw a network that will meet the Highway Department's requirements. What is the total number of miles that must be plowed?

b. As you know from this lesson, there may be several networks that satisfy the Highway Department's requirements. Find all the plowed-road networks that will work. Check the total length of each such network and make sure that you get the same total mileage for each that you got in Part a.

c. Construct a distance matrix for each shortest network (that is, for each minimal spanning tree) that you found in Parts a and b.

d. For each plowed-road network, which town is most centrally located? On what quantitative (numerical) information did you base your decision?

e. Each plowed-road network that you found has the same total length. Despite this, do you think one is better than another? Justify your answer by using information from the graphs and the distance matrices.

2. A family with seven members in different parts of the country has a relative working overseas. The family wants to set up a telephone-calling network so everyone will know the latest news about the overseas relative, for the least total cost. The relative overseas will call Felix, and then Felix will start the message through the network. The table below shows the cost for a 15-minute phone call between each pair of family members.

Phone Call Costs

	Amy	Felix	Hillary	Kit	Owen	Pearl	Robin
Amy		$3.50	$4.75	$3.80	$4.10	$2.85	$5.10
Felix	$3.50		$3.75	$2.50	$4.50	$4.10	$3.40
Hillary	$4.75	$3.75		$2.95	$3.15	$4.40	$3.50
Kit	$3.80	$2.50	$2.95		$4.25	$3.30	$3.40
Owen	$4.10	$4.50	$3.15	$4.25		$2.95	$3.25
Pearl	$2.85	$4.10	$4.40	$3.30	$2.95		$3.60
Robin	$5.10	$3.40	$3.50	$3.40	$3.25	$3.60	

a. What is the total cost of the least expensive calling network they can set up?

b. Write a description of who should call whom in this least expensive calling network.

3. There are many situations in which it is useful to detect *clustering*. For example, health officials might want to know if outbreaks of the flu are spread randomly over the country or if there are geographic clusters where high percentages of people are sick. Geologists might want to know if the distribution of iron ore is spread evenly through an ore field or if high densities of ore are clustered in particular areas. Economists might want to know if small business start-ups are more common, that is, clustered, in some areas. There are several techniques that have been devised to detect clustering. A technique involving minimal spanning trees is illustrated in the following copper-ore mining context.

Great Lakes Mining Company would like to know if copper ore is evenly distributed throughout a particular region or if there are clusters of ore. The company drills a grid of nine test holes in each of two ore fields. The following diagrams show the grid of test holes in each ore field, along with the percentage of copper, expressed as a decimal, in the sample from each test hole.

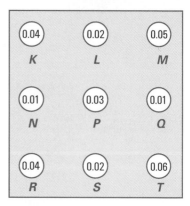

a. Construct a graph model for the grid in each ore field. Represent each test hole as a vertex, and connect two vertices with an edge if the test holes are next to each other (vertically, horizontally, or diagonally).

b. For each edge, compute the absolute value of the difference between the concentrations of copper at the two vertices on the edge. Label the edge with this number. For example, consider the grid on the left. Since the concentration at test hole A is 0.01 and the concentration at test hole E is 0.04, label the edge connecting A and E with $|0.01 - 0.04|$ or 0.03.

c. Find a minimal spanning tree for each of the two graphs. What is the total length of each minimal spanning tree? (Note that "total length" here refers to the sum of the concentration differences on each edge.)

d. Now consider the connection between the length of a minimal spanning tree and clusters of ore concentrations.

- In general, if there is a cluster of test holes with similar concentrations of copper, will the numbers on the edges in that cluster be large or small? Why?

- If there is more clustering in one of the ore fields, will the length of the minimal spanning tree for that ore field be larger or smaller than the other one? Why?

- Which of the two ore fields in this example has greater clustering of concentrations of copper? Explain in terms of minimal spanning trees.

4. A restaurant has opened an outdoor patio for evening dining. The owner wants to hang nine decorative light fixtures at designated locations on the overhead latticework. Because of the layout of the patio and the latticework, it is not possible to install wiring between every pair of lights. The matrix below shows the distances in feet between lights that can be linked directly. The main power supply from the restaurant building is at location X. The owner wants to use the minimum amount of wire to get all nine lights connected.

	X	A	B	C	D	E	F	G	H	I
X	–	18	–	–	11	–	–	13	17	–
A	18	–	16	–	–	15	15	–	–	–
B	–	16	–	16	12	–	–	–	–	–
C	–	–	16	–	–	–	–	12	–	–
D	11	–	12	–	–	–	–	10	–	–
E	–	15	–	–	–	–	7	–	–	–
F	–	15	–	–	–	7	–	–	–	–
G	13	–	–	12	10	–	–	–	18	–
H	17	–	–	–	–	–	–	18	–	8
I	–	–	–	–	–	–	–	–	8	–

a. What is the minimum amount of wire needed to connect all nine lights?

b. Suppose the electrician decides to start at the power supply, X, then go to the closest light, then go to the closest light from there, and so on. What algorithm does she seem to be using?

c. Apply the electrician's algorithm to the graph, starting at X. Describe what happens.

Organizing

1. In this lesson, you found shortest networks by finding minimal spanning trees. Graphs for problems involving minimal spanning trees always have numbers associated with their edges. Now consider graphs that do not have numbers on the edges. In these cases, you might be interested in finding a network that connects all the vertices of a graph and uses the fewest number of edges. Such a graph is called a **spanning tree**.

   a. Find a spanning tree for each graph below. Describe the method you used to find the spanning trees.

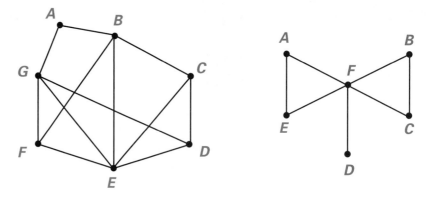

   b. Find three different spanning trees for the following graph.

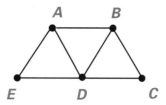

   c. Write a rule relating the number of vertices in a graph and the number of edges in a spanning tree for the graph.

2. Recall that a tree is a connected graph that has no circuits. In Course 1, you investigated vertex colorings of graphs. In this task, you will consider vertex colorings of trees.

   a. Draw three different trees, each having at least six vertices.

   b. What is the minimum number of colors needed to color the vertices of each tree so that any two vertices connected by an edge have different colors?

   c. What is the minimum number of colors needed to color the vertices of *any* tree so that two vertices connected by an edge have different colors? Explain your reasoning.

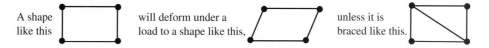

3. If you make a rectangular frame, like framing used for scaffolding, it is necessary to brace it with a diagonal strip. Without such a strip it can deform, as illustrated below. (It doesn't matter which diagonal is used.)

A shape like this  will deform under a load to a shape like this,  unless it is braced like this.

Buildings and bridges often are constructed of rectangular steel grids, such as those shown below. To make grids rigid, you do not have to brace each cell with a diagonal, but you do have to brace some of them.

a. One of the two grids below is rigid, and the other is not.

Grid A   **Grid B**

- Which is the rigid grid? Explain your choice.
- For the rigid grid, remove some of the braces without making the grid nonrigid. (You may want to make a physical model to help.)
- For the nonrigid grid, add some diagonal braces to make the grid rigid.

b. You can use vertex-edge graphs to help solve problems like those above. The first step is to model the grid with a graph. Let the vertices represent the rows and columns of the grid. Then draw an edge between a row-vertex and a column-vertex if the cell for that row and column is braced. The graph for Grid A is drawn below.

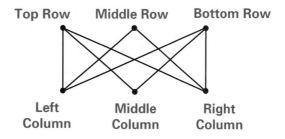

- Explain why there is an edge from the top-row vertex to the middle-column vertex.
- Explain why there is no edge from the middle-row vertex to the middle-column vertex.
- Construct the graph for Grid B.

c. Compare the graphs for Grid A and Grid B.

  ■ Is the graph for Grid A connected? Is Grid A rigid?

  ■ Is the graph for Grid B connected? Is Grid B rigid?

  ■ Do you think connectedness of a graph model of a grid will ensure the corresponding grid is rigid? Draw another connected graph and another nonconnected graph, each of which could model a grid. What is true about the rigidity of their corresponding grids? Do these examples support your conjecture?

d. A rigid grid may have "extra" bracings. For example, in Part a you discovered that it was possible to remove some of the cell bracings of Grid A and still maintain the rigidity. Now investigate what this means in terms of the corresponding graph.

  ■ On a copy of the graph for Grid A, eliminate "extra" edges, one at a time. Stop when you think that removing another edge will result in a graph that represents a nonrigid grid. How is your final "subgraph" related to the original graph?

  ■ Draw a rigid grid that has the minimum number of bracings. Examine its corresponding vertex-edge graph representation. Is there anything special about this graph?

  ■ What feature of a graph model of a rigid grid would indicate that the grid has the minimum number of bracings? Make a conjecture. Add the minimum number of bracings to Grid B to make it rigid, and then examine the corresponding graph. Does this example support your conjecture?

4. In Organizing Task 3, you discovered how to use graphs to analyze rigidity of rectangular grids. You probably found the following:

  ■ A grid is rigid provided its graph is connected.

  ■ The minimum number of bracings to make the grid rigid have been used only when its graph is a spanning tree.

Consider the grid shown at the right.

a. Draw the graph for this grid.

b. Is the grid rigid? How can you tell by looking at the graph?

c. Have the minimum number of cells been braced? How can you tell by looking at the graph?

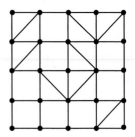

d. Remove "extra" bracings by removing edges that create circuits.

e. Draw the corresponding grid with a minimum number of bracings to make it rigid.

f. The key fact about grids and graphs is that a grid is rigid provided its graph is connected. Explain why this fact makes sense.

Reflecting

1. Refer back to Organizing Task 1 on page 332. Graphs without numbers on their edges can be considered graphs *with* numbers on the edges by using the same number on all the edges. For example, you could put a "1" on each edge. (The number on an edge is called its *weight*. You will further investigate weighted graphs in Lesson 2.)

   a. Explain why putting the same number on each edge is like having no numbers on the edges.

   b. What does this suggest about the results and algorithms created for graphs with numbers on their edges?

2. Think about the characteristics of those graphs which are also trees.

   a. Explain why a tree can be considered a *minimal* connected graph.

   b. Explain why a tree can be considered a *maximal* graph with no circuits.

3. There is a story that when composer Igor Stravinsky (1882–1971) was asked how he would describe his music pictorially, he replied, "This is my music:"

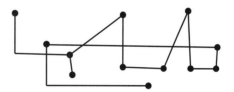

   a. What do you think Stravinsky meant?

   b. One of Stravinsky's most well-known compositions is *The Rite of Spring*. Listen to *The Rite of Spring*. Why do you think Stravinsky used a tree graph to describe his music? (Incidentally, because *The Rite of Spring* sounded so unexpectedly different and unusual, the premiere performance caused a riot in the audience.)

4. Think of a situation, different from any in this lesson, in which it would be helpful to find a minimal spanning tree. Describe the situation and explain how a minimal spanning tree could be used.

Extending

1. There is a popular solitaire game called *Clock Solitaire*. As with many games, there is a lot of mathematics underlying the game. In fact, with a little vertex-edge graph modeling, you can determine whether you will win the game before you actually start playing.

Here's how to play the game:

Start with a standard deck of 52 cards and make 13 piles of 4 cards each, with all cards face down. Place 12 of the piles in the positions of the numbers on a clock face, and place the 13th pile in the center of the clock. Turn over the top card of the center pile and slide it face-up under the pile whose number corresponds to the face value of the card. (Aces go to 1:00, jacks to 11:00, queens to 12:00, and kings go to the center pile. For example, if you turn over the top card of the center pile and it is a nine, then you would slide it face-up under the pile at 9:00.) Then turn over the top card on that pile, and place the new card where it indicates. You continue in this manner until it is no longer possible to turn up a card on the pile where you are. You win if you have turned up all 52 cards.

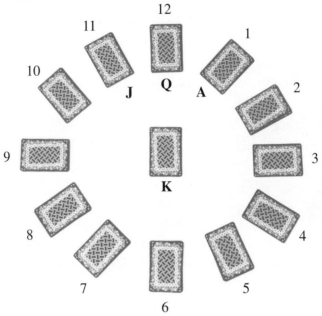

Get a deck of cards and play the game a few times. Then think about some patterns in the game.

a. After playing the game a few times, you will notice that the last play of the game is always on the center pile. Explain why this happens.

b. Set up another game of *Clock Solitaire,* but don't begin playing. Construct a graph model as follows, and call it the *total graph* for your game. Let the vertices be the positions of the piles. For each vertex, draw a directed edge from the vertex to each of the vertices indicated by the cards in the pile. So, for example, if the pile at 3:00 has a 9, 3, 2, and jack in it, then you would draw directed edges from the vertex at 3 to the vertices at 9, 3, 2, and 11. (The edge from 3 to 3 would be a loop.)

- How many directed edges will be coming into each vertex? Why?

- Explain why winning the game means that there is an Euler circuit in the total graph. (Recall that an Euler circuit is a circuit through a graph that uses each edge exactly once.)

c. The total graph is messy because it has so many edges. A simpler graph model, call it the *bottom-card graph*, is constructed as follows. Once again let the vertices be the positions of the piles. For every pile except the center pile, look at the *bottom* card and draw a directed edge from the vertex for that pile to the vertex indicated by the bottom card. For example, if the bottom card of the pile at 5:00 is a queen, then you would draw a directed edge from the vertex 5 to the vertex 12.

- Set up a game of *Clock Solitaire* and construct the bottom-card graph for the game.

- Because of the rules for how it is constructed, a bottom-card graph never has an edge coming out of the center vertex, where the kings go. If you win the game, will there be an edge going into the center vertex? Explain.

d. Here is an amazing fact: *You win the game exactly when the bottom-card graph is a tree.*

- Set up two games of *Clock Solitaire*, one with a bottom-card graph that is a tree and one with a graph that isn't. Play the games and see if you win one and lose the other.

- *Challenge*: Write a paragraph explaining why the amazing fact is true. **Hints:** Think of each edge on the bottom-card graph as representing the "last exit" out of the pile. Think of the bottom-card graph as part of the total graph.

2. The nearest-neighbor algorithm you investigated in Activity 9 on page 324 did not always produce a minimal spanning tree in a connected graph. Below is a modified version of that algorithm, called *Prim's algorithm*.

    i. Make a copy of the graph with the edges drawn lightly.

    ii. Choose a starting vertex. This is the beginning of the tree.

    iii. Find all edges that have one vertex in the tree constructed so far. Darken the shortest such edge that does not create a circuit. If there is more than one such edge, choose any one.

    iv. Repeat Step iii until all vertices have been reached.

a. Test Prim's algorithm using three copies of the following graph from Activity 9.

- Apply the algorithm starting at vertex *E*.

- Apply the algorithm starting at vertex *C*.

- Apply the algorithm starting at vertex *A*.

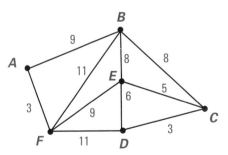

b. Compare Prim's algorithm with the nearest-neighbor algorithm in Activity 9 on page 324.

■ How are they similar? How are they different?

■ Compare the results in Part a above to those you got with the nearest-neighbor algorithm in Activity 9.

c. Explain why a tree is constructed at each stage of Prim's algorithm.

d. Do you think Prim's algorithm is a good procedure for finding a minimal spanning tree? Write a brief justification of your answer.

e. Compare Prim's algorithm with Kruskal's best-edge algorithm from Activity 7 on page 323.

■ How are they similar? How are they different?

■ Which algorithm would you prefer to use to find a minimal spanning tree? Why?

f. To find minimal spanning trees for large graphs, algorithms such as Prim's or Kruskal's must be implemented on a computer. Do you think it might be the case that the easier (preferred) algorithm for finding a minimal spanning tree for small graphs by hand is different from the best (most efficient) algorithm for large graphs by computer? Explain your reasoning.

Joseph Kruskal

3. To find a minimal spanning tree in a graph, you look for a network of existing edges that joins all the vertices and has minimum length. In some situations, you may want to create a minimal spanning network by adding new vertices and edges to the original graph. Such a network is called a **Steiner tree** (named after Jacob Steiner, a 19th century mathematician at the University of Berlin).

There is a way to use geometry to find Steiner trees. For this task, you should use a geometry software package that allows you to construct, measure, and move geometric figures.

a. Using a geometry software package, construct a triangle in which all the angles are less than 120°. (Actually, because of software limitations, it is best if all angles are less than 115°.) You can consider this triangle as a vertex-edge graph.

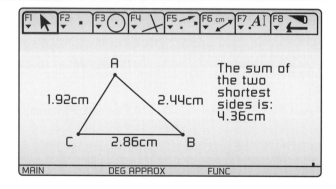

Your goal in Parts b and c is to find a shortest network that joins all three of the triangle's vertices.

b. A minimal spanning tree for this triangular graph is just the network consisting of the two shortest sides. Find the sum of the lengths of the two shortest sides. Now you have the length of a minimal spanning tree.

c. Now consider a network where you are allowed to insert a new vertex and new edges. You will investigate whether this gives you a connected network that is shorter than the minimal spanning tree you found in Part b. Begin by using the software to perform the following construction:

- Insert a new point (vertex) inside the triangle.

- Construct segments from the inside point to each of the vertices of the triangle. (This gives you a network that connects all three of the triangle's vertices, but it uses a vertex and edges that are not part of the original triangle.)

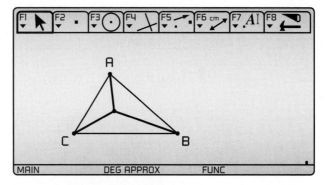

- Measure the length of this network by measuring each segment and adding the three lengths. (For best results, set the measurement precision of your geometry software so that lengths are measured in tenths of pixels.)

- Now use the software to grab the inside point and drag it around. Note that the network length changes as the point is moved. Drag the point around until the network length is as small as possible.

Is this length smaller than the minimal spanning tree length from Part b?

d. Find the measure of the three central angles that surround the inside point. (For best results, set the measurement precision of your geometry software to measure angles in whole-unit degrees.) Make a conjecture about the measures of these angles when the inside point is moved to a position giving the shortest connected network. Test your conjecture on some other triangles.

4. Write a calculator or computer program for the best-edge algorithm. Use your program to find a minimal spanning tree for the graphs in Modeling Task 1 on page 328 and Extending Task 2 on page 337.

Lesson 2 *Shortest Paths and Circuits*

In the last lesson, you investigated minimal spanning trees and their applications to finding shortest networks. The graphs you used to model situations had numbers associated with their edges. The numbers most often represented distance, but sometimes they represented other quantities like cost or concentration of copper ore. A graph with numbers on its edges is called a **weighted graph**, and the numbers, whatever they represent, are called **weights**.

In this lesson, you will investigate paths and circuits in weighted graphs. Problems of this type are common in the communications, shipping, and travel industries.

Think About This Situation

Rely on your common-sense notion of shortest paths and shortest circuits as you try to answer the following questions.

a What kinds of situations in the shipping or travel industry might involve shortest paths or shortest circuits?

b What kinds of situations in the communications industry might involve finding a shortest path or shortest circuit?

INVESTIGATION 1 ▶ Shortest Routes

Shortest routes are important in many different contexts. Several are explored in this investigation.

Airfares Getting the least expensive airfare these days is not a simple matter. The fare you pay depends on many factors, such as how far in advance you buy your ticket or if you will stay over Saturday night. It even depends on the cities through which you fly to get to your final destination.

Kansas City, Kansas

Detroit

1. The diagram below shows sample airfares on a major airline for round-trip tickets, purchased two weeks in advance, with a Saturday night stay-over.

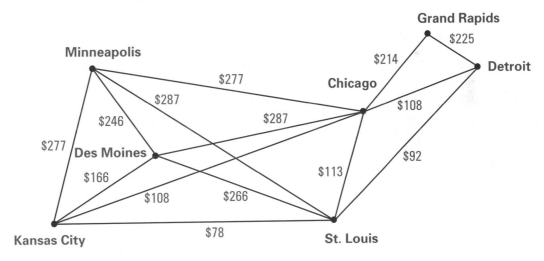

a. Thinking of this diagram as a weighted graph, what does the shortest path between two vertices tell you about airfare?

b. What is the cheapest airfare between Minneapolis and Detroit? Between Kansas City and Grand Rapids?

c. Why do you think Chicago and St. Louis are called major "hubs"?

d. If a friend was planning to travel from one of the three westernmost cities to Detroit, what advice would you give him or her?

Manufacturing Shortest paths also can be used to optimize a manufacturing process. Consider a toy company that makes a hand-crafted game which involves moving pegs around a wooden game board. In the following activity, you will analyze how the game board is made.

2. Three of the tasks required to make the game board are cutting the wood (*C*), drilling the holes (*H*), and stripping the wood (*S*). These three tasks can be done in any order, but the time required to do each task depends on when it is done. The manufacturer wants to answer the following question:

   What is the most efficient order for doing the three tasks?

   All the information about the tasks, their orders, and their times (in minutes) is given in the digraph below.

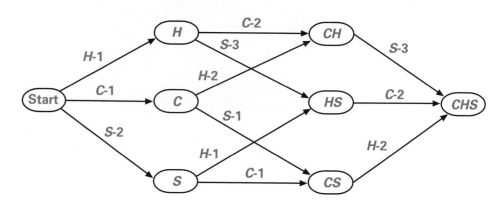

The vertices of this digraph represent stages of the manufacturing process. For example, "*S*" represents a piece of wood that has been stripped only; "*CS*" represents a piece of wood that has been cut and stripped but has not yet had holes drilled in it.

The labels on the edges tell which task will lead to the next stage and how many minutes that task will take. Remember that the time required for a task depends on when it is done. Thus, for example, the "*C*-1" label on the edge from *S* to *CS* means that it takes 1 minute to cut a piece of wood that has been stripped, while the "*C*-2" label on the edge from *H* to *CH* means that it takes 2 minutes to cut a piece of wood that has holes drilled in it.

a. Begin your analysis by practicing reading the digraph.

■ Which vertex represents the stage in the manufacturing process where a piece of wood has been stripped and has holes drilled in it, but the wood has not yet been cut?

■ What does the vertex "*CHS*" represent?

■ How long does it take to strip a piece of wood that has been cut and has holes drilled into it?

■ What does the label "*S*-1" on the edge from *C* to *CS* mean?

b. Now interpret paths in the digraph.

- What order of tasks is represented by the path: Start→*S*→*CS*→*CHS*? What is the length, in minutes, of this path?

- What does the length of a path tell you about the manufacturing process for the game board?

- How many different ways can you order the three tasks: *C*, *H*, and *S*? How does this relate to the number of different paths from Start to *CHS*?

c. Explain how to use paths in the digraph to answer these questions.

- What is the most efficient order for doing the three tasks?

- What is the total time required to manufacture the game board if the tasks are done in this optimal order?

Road Networks Reproduced below is the weighted graph representing the network of seven rural towns you investigated in Lesson 1. (The weights on the edges represent distances in miles). In that lesson, you found a minimal spanning tree that represented a paved-road network connecting all the towns. Only some roads were paved, and there was only one path from a given town to any other town on the paved-road network.

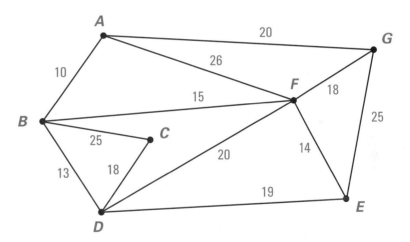

3. For this new situation, assume that all the roads are paved and consider all the possible paths between each pair of towns.

a. Find three different paths from *A* to *E*. List the towns that make up each path.

b. What is the shortest path from *A* to *E*? What is the length of the shortest path?

c. What is the shortest path from *F* to *A*? What is its length?

d. A distance matrix shows the length of a *shortest* path between every two towns in a network. Finding all the shortest paths in this network is a lot of work since there are many paths between each pair of towns. An efficient algorithm for finding shortest paths is described in Extending Task 2, page 359, but it is very time-consuming to carry out by hand. The SHORTCUT software developed for your calculator, or similar computer software, can implement this algorithm for you. Use software to construct the distance matrix for the road network on the previous page.

e. Examine the distance matrix displayed by the software.

- Compare the matrix entry for the shortest distance between *A* and *E* with what you found in Part b.

- What is the farthest you would have to drive to get from one town to any other town?

- For each town, compare its row to its column. What pattern do you see? Explain why this pattern should be expected.

f. Now consider the row sums that the SHORTCUT program, or a similar program, computed.

- The row sum for vertex *F* is less than the row sum for vertex *G*. What does this mean in terms of towns and distances?

- Which town is the most isolated?

- Suppose a new county hospital to serve all seven towns is to be built in one of the towns. In which town should it be built? Why?

Checkpoint

Look back at your work finding shortest paths and consider how such paths differ from minimal spanning trees.

ⓐ What information did shortest paths give you in each of the three contexts in this investigation?

ⓑ What is the difference between a minimal spanning tree and a shortest path?

ⓒ Describe a context different from those in Lessons 1 and 2 in which you would want to find a minimal spanning tree.

ⓓ Describe a new context in which you would want to find a shortest path.

Be prepared to share the results of your analyses with the class.

▶ **On Your Own**

Consider the road network below, where distances are shown in miles.

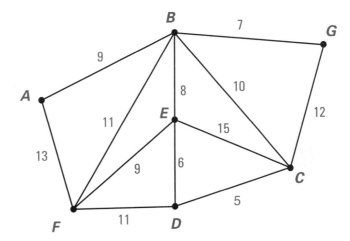

a. Suppose some of the roads need to be plowed after a snowstorm. Find the shortest possible network of plowed roads that will allow cars to drive from every town to every other town on plowed roads.

b. What is the shortest distance from *A* to *F* on the plowed-road network?

c. When there's no snow, all the roads can be used. Find the shortest distance from *A* to *F* when all the roads are clear.

INVESTIGATION ▶ 2 Graph Games

Graphs are not only helpful in solving practical problems, they are also great for games. Described below are two classic games based on the same graph model.

1. Sir William Rowan Hamilton, a famous Irish mathematician, invented a game in 1857 based on a dodecahedron. Recall that a dodecahedron has 20 vertices and 12 faces, which are regular pentagons. The *Traveler's Dodecahedron* was a game consisting of a wooden dodecahedron with a peg at each vertex, and some string. The vertices were labeled with the names of cities from around the world, like Canton, New Delhi, and Zanzibar. The object of the game was to start at one city, visit the other 19 cities *only once*, and end back where you started. The string was wound around the pegs to keep track of the journey.

Since the dodecahedron was difficult to carry around, Hamilton made another version of the game, called the *Icosian Game*. The Icosian Game was made by "flattening" the dodecahedron into a vertex-edge graph with 20 vertices and 11 pentagonal regions.

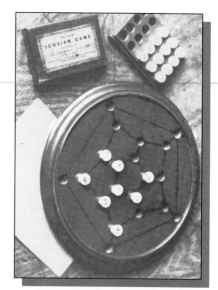

a. Draw a vertex-edge graph representing the Icosian Game.

b. Instead of using string, use your pencil to trace a path that will win the game.

c. Does it matter where you start? Explain why or why not.

A path like the one you traced to win the Icosian Game is called a Hamiltonian circuit. A **Hamiltonian circuit** is a circuit through a graph that starts at one vertex, visits all the other vertices *exactly once*, and finishes where it started. (By the way, Hamilton's games were interesting mathematically, but they were commercial disasters.)

2. A well-known game among chess players is called the *Knight's Tour Problem*. In one version of this game, you start with a knight on any square of a chessboard and try to visit each of the other squares exactly once by successively moving the knight. You must finish the knight's tour where you began.

A knight moves in an L-shaped pattern. For example, four sample moves of a knight are illustrated below. Any such L-shaped move is a legitimate move.

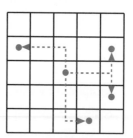

a. Explain how to model and solve this game using a graph and a Hamiltonian circuit. Don't solve the game yet, just explain how you could solve it using a Hamiltonian circuit.

b. Consider the modified chessboard below. The square with a cross in it is a special part of the chessboard; you can cross it, but you cannot land on it.

A	B	C
D	✕	E
F	G	H

Construct a graph model for the knight's tour problem on this modified chessboard. (Put lots of space between vertices so that the graph will be easy to read.) Solve the problem by finding a Hamiltonian circuit, if possible. Explain your solution.

c. Use a Hamiltonian circuit to find a knight's tour on the "chessboard" below, if possible.

I	J	K
A	B	C
D	✕	E
F	G	H

3. Some graphs have Hamiltonian circuits and some do not. Unlike the case of Euler circuits, no one has yet found a simple test for determining whether or not a graph has a Hamiltonian circuit. Can you think of a property of a graph that will guarantee that it does *not* have a Hamiltonian circuit? Make a conjecture and defend it.

Checkpoint

Recall from Course 1 that an Euler circuit is a circuit that uses each edge exactly once. You now have studied another kind of circuit in vertex-edge graphs—a Hamiltonian circuit.

ⓐ What is the difference between an Euler circuit and a Hamiltonian circuit?

ⓑ Consider this graph:

- Find an Euler circuit.
- Find a Hamiltonian circuit.

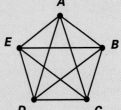

Be prepared to share your thinking and circuits with the entire class.

On Your Own

The Checkpoint on page 347 gives an example of a graph that has both an Euler circuit and a Hamiltonian circuit. Draw the following graphs (if possible).

a. The graph does not have an Euler circuit or a Hamiltonian circuit.

b. The graph does not have an Euler circuit but does have a Hamiltonian circuit.

c. The graph has an Euler circuit but does not have a Hamiltonian circuit.

INVESTIGATION 3 The Traveling Salesperson Problem

The *Traveling Salesperson Problem* is one of the most famous problems in mathematics. It can be thought of as another game, but, as you will see, it also has many important applications. Here's the problem:

> *A sales representative wants to visit several different cities, each exactly once, and then return home. Among the possible routes, which one will minimize the total distance traveled?*

1. Consider the Traveling Salesperson Problem in the context of this airfare graph.

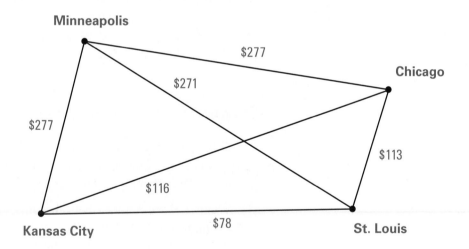

a. Without help from others, solve the Traveling Salesperson Problem for this weighted graph. That is, find a circuit that visits each of the cities exactly once and has the minimum total weight. What does "weight" represent in this case?

b. Compare your solution to those of other students. Resolve any differences.

c. Now write a description of a method for finding the optimal circuit.

d. How do you know that there is no circuit less expensive?

e. How many different Hamiltonian circuits are there in this graph? For the purpose of finding the total cost of circuits, two circuits are different only if they have different edges. It doesn't matter where you start or which direction you go around the circuit.

f. Could you generalize your method in Part c to find the optimal circuit for traveling to the capitals of all 48 contiguous states? Explain your reasoning.

2. Compare solving the Traveling Salesperson Problem to some of the other problems you have solved in this unit.

a. What is the relationship between a Hamiltonian circuit and a solution to the Traveling Salesperson Problem?

b. Describe how a solution to the Traveling Salesperson Problem is similar to, and yet different from, a minimal spanning tree.

c. This investigation is part of a lesson entitled "Shortest Paths and Circuits." Explain how a solution to the Traveling Salesperson Problem is a shortest path or circuit.

3. In Lesson 1, you used Kruskal's best-edge algorithm to find a minimal spanning tree. One group of students devised the following best-edge algorithm for the Traveling Salesperson Problem. They claim it will solve the Traveling Salesperson Problem.

   i. Make a copy of the graph with the edges drawn lightly.

   ii. Darken the shortest edge not yet used, provided that:

   ■ you do not create a circuit of darkened edges, unless all the vertices are included;

   ■ no vertex is touched by three darkened edges.

   (The edge you darken does not have to be connected to previously darkened edges.)

   iii. Repeat Step ii as long as it is possible to do so.

a. Analyze this algorithm.

   ■ Why do you think this algorithm requires that you do not create a circuit of darkened edges, unless all the vertices are included?

   ■ Why do you think this algorithm requires that no vertex is touched by three darkened edges?

b. Apply the algorithm to the airfare graph in Activity 1, page 348. Does this algorithm produce a solution to the Traveling Salesperson Problem?

4. One method that certainly will work to solve the Traveling Salesperson Problem is to list all possible circuits, compute the length of each one, and choose the shortest. This approach of checking all possibilities is sometimes called a **brute force method**. With computers available to do all the calculations, you might think this is the way to proceed.

However, think about how long it will take a computer to use the brute force method to solve the Traveling Salesperson Problem for the 26 cities shown in the following map. Assume each city is connected directly to all the others, and the tour starts at Atlanta.

a. Starting from Atlanta, how many cities could be the first stop?

b. Once you choose a city for this first stop, how many cities could be the second stop in the circuit? Remember that every city is connected directly to every other city, and each city is visited exactly once.

c. How many different first-stop/second-stop combinations are there? Justify your answer.

d. How many cities could be the third stop of the circuit? How many different combinations of first-stop/second-stop/third-stop are there?

e. How many different circuits are possible using all the cities?

f. Suppose a computer program can compute the length of one billion circuits per second. How many seconds will it take to compute the length of all the circuits? How many years?

g. Under what conditions do you think the brute force method is a practical way to solve the Traveling Salesperson Problem?

You have now investigated several new graph models, along with associated algorithms and the so-called brute force method.

a In Investigations 2 and 3, you used Hamiltonian circuits to model several different situations. For each situation, describe what the vertices, the edges, and a Hamiltonian circuit represent.

b A best-edge algorithm worked to find minimal spanning trees. Can you always solve the Traveling Salesperson Problem using a best-edge algorithm?

c Brute force methods can be used to solve many graph problems.

- Describe the brute force method for solving the Traveling Salesperson Problem.

- Describe brute force methods for finding shortest paths and minimal spanning trees.

- Why aren't brute force methods practical for large graphs?

Be prepared to share your descriptions and thinking with the class.

In this lesson, you have explored two famous problems in mathematics: the general Traveling Salesperson Problem and characterizing graphs that have a Hamiltonian circuit. Both of these problems are currently unsolved! New applications and new mathematics have been developed as researchers continue to work on these problems.

For the Traveling Salesperson Problem, the goal is to find an efficient solution that will work in all situations. You have seen one method, the best-edge algorithm, that is efficient but does not guarantee a solution. You have seen another method, the brute force method, that guarantees a solution but is not efficient. No one knows a method that is both efficient and works in all situations.

Since solving the Traveling Salesperson Problem has applications for so many different kinds of networks, like telephone and transportation networks, mathematicians are always looking for better algorithms that will solve the problem for larger graphs. In 1986, the Traveling Salesperson Problem was solved efficiently for a graph with 532 vertices, and by 1994 the record was 7,397 vertices. More recently, mathematicians have found methods that are guaranteed to find a circuit that is less than one percent longer than the shortest possible circuit, even for graphs with several hundred thousand vertices.

The matrix below shows the mileage between four cities.

$$
\begin{array}{c}
 \\
A \\
B \\
C \\
D
\end{array}
\begin{array}{cccc}
A & B & C & D \\
\left[\begin{array}{cccc}
0 & 20 & 25 & 40 \\
20 & 0 & 35 & 45 \\
25 & 35 & 0 & 30 \\
40 & 45 & 30 & 0
\end{array}\right]
\end{array}
$$

a. Represent the information in the matrix with a weighted graph.

b. Trace all the different Hamiltonian circuits starting at A. List the vertices in each circuit.

c. Record the total length of each circuit.

d. Would you get different answers in Part c if the starting vertex was B?

e. Is there a difference between circuit A-B-C-D-A and circuit A-D-C-B-A? Explain your reasoning.

f. What is the solution to the Traveling Salesperson Problem for this graph?

MORE
Modeling • Organizing • Reflecting • Extending

Modeling

1. Integrated circuit boards are used in a variety of electronic devices, including modern kitchen appliances, video games, automobile ignition systems, and the guidance systems in commercial airliners. To manufacture a circuit board, a laser must drill as many as several million holes on a single board. This usually is done with a laser in a fixed position; the circuit board is turned to the positions that must be drilled. For maximum efficiency, the board must end up in its original position, no hole should pass under the laser more than once, and the total distance that the board is moved should be as small as possible.

To see how this problem is solved using graphs, consider a simple situation in which there are just four holes to be drilled. The distance, in millimeters, that the board must be moved from one hole to another is given in the matrix below.

$$
\begin{array}{c} \\ A \\ B \\ C \\ D \end{array}
\begin{array}{cccc}
A & B & C & D \\
\left[\begin{array}{cccc}
- & 0.02 & 0.02 & 0.01 \\
0.02 & - & 0.04 & 0.02 \\
0.02 & 0.04 & - & 0.05 \\
0.01 & 0.02 & 0.05 & -
\end{array}\right]
\end{array}
$$

a. Represent the information in the matrix with a weighted graph.

b. Explain why solving the circuit board problem is the same as solving the Traveling Salesperson Problem for this graph.

c. Find the order for drilling the holes that will minimize the total distance that the board has to be moved.

2. Information is transmitted between computers by converting the information into strings of 1s and 0s and then sending these strings as electronic signals from one computer to another. The method used to translate the information into 1s and 0s is called a *code*. The computers that send and receive the information know how to create and interpret the code.

One commonly used code is called a *Gray Code*. A Gray Code is a list of 0-1 strings with the following properties:

- Every string of a given length is in the list.
- Each string in the list differs from the preceding one in exactly one position.
- The first and last strings in the list differ in exactly one position.

a. Here is a Gray Code using strings of length two:

10   00   01   11

Verify that the three properties of a Gray Code are satisfied.

When the strings are short and there are so few of them, it is possible to find a Gray Code by trial and error. Using Hamiltonian circuits is one way to find Gray Codes with longer strings. Consider strings of length three.

b. A string of length three has a 1 or a 0 in each of the three positions. For example, 100 and 011 are 0-1 strings of length three. List all eight 0-1 strings of length three.

c. Build a graph model by letting the vertices be the eight 0-1 strings and connecting two vertices with an edge if the two strings differ in exactly one position.

Cray Supercomputer

d. Find a Hamiltonian circuit in the graph and then list all the vertices in the circuit in order.

e. Is the list you made in Part d a Gray Code? Why or why not?

f. Find another Gray Code using strings of length three.

3. A company is expanding into a new region of the country. It will set up offices in five cities in the region. The airfare (in dollars) for direct flights between each pair of cities is shown in the matrix below.

	A	B	C	D	E
A	–	500	400	250	100
B	500	–	200	900	250
C	400	200	–	100	250
D	250	900	100	–	550
E	100	250	250	550	–

One of the five cities will be the regional headquarters, at which regular meetings will be held. The president of the company asks you to use the information in the matrix to recommend one city to be the regional headquarters. Make a recommendation and defend your choice.

4. Three different features appear in a local newspaper every day. The features are scheduled to be printed in three jobs, all on the same printing press. After each job, the press must be cleaned and reset for the next job. After the last job, the press is reset for the first job to be run the next morning. The time, in minutes, needed to set up the press between each pair of jobs is shown in the matrix below.

	A	B	C
A	–	25	15
B	30	–	25
C	20	20	–

The newspaper production manager wants to schedule the jobs so that the total set-up time is minimal.

a. Model this situation with a *weighted digraph*.

b. Show how a solution to the Traveling Salesperson Problem will tell you how to schedule the jobs so that the total set-up time is minimal.

c. In what order should the jobs be scheduled and what is the minimum total set-up time?

5. Martin and William have invited friends, who do not all know each other, for dinner. All eight people will be seated at a round table, and the hosts want to seat them so that each guest will know the people sitting on each side of him or her. In the graph below, two people who know each other are connected by an edge.

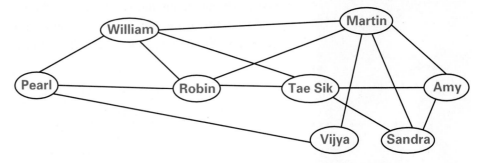

   a. Show how you can use a Hamiltonian circuit to decide how to seat the people according to Martin and William's requirement.

   b. Sketch a diagram of the round table and show how the people should be seated.

Organizing

1. A key characteristic of vertex-edge graphs is that the position of the vertices and the actual geometric length of the edges do not matter. All that matters is the way in which the edges connect the vertices. So, for example, when drawing a graph that represents a distance matrix, the graph does not have to be drawn to scale. In geometry, on the other hand, position and length are important factors. To appreciate this important difference between geometry and graph theory, consider the following shortest distance matrix. Each entry shows the shortest distance, in miles, between two corresponding towns.

$$\begin{array}{cc} & \begin{array}{ccc} W & R & T \end{array} \\ \begin{array}{c} \text{Woebegone (W)} \\ \text{Rivendell (R)} \\ \text{Troy (T)} \end{array} & \left[\begin{array}{ccc} - & 60 & 100 \\ 60 & - & 80 \\ 100 & 80 & - \end{array}\right] \end{array}$$

   a. Draw a vertex-edge graph that represents the information in the matrix.

   b. Use a compass and ruler to draw a scale diagram showing the distances between the three towns. Assume straight-line roads between the towns.

   c. State a question involving these three towns that is best answered using a geometric model.

   d. State a question that could be answered using either model.

2. A **Hamiltonian path** is a path that visits each vertex of a graph exactly once. Thus, a Hamiltonian circuit is a special type of Hamiltonion path: one which starts and ends at the same vertex. Hamiltonian paths can be used to analyze tournament rankings.

Consider a round-robin tennis tournament involving four players. The matrix below shows the results of the tournament. Recall that the matrix is read from row to column, with a "1" indicating a win. For example, the "1" in the Flavio-Simon entry means that Flavio beat Simon.

Tournament Results

	J	S	F	B
Josh (J)	0	0	0	1
Simon (S)	1	0	0	1
Flavio (F)	1	1	0	1
Bill (B)	0	0	0	0

a. Draw a digraph representing the information in the matrix.

b. Find all the Hamiltonian paths in the graph.

c. Use the Hamiltonian path to rank the players in the tournament. Explain the connection between your ranking and the Hamiltonian path.

d. In the "Matrix Models" unit of Part A of this course, you used row sums and powers of matrices to rank tournaments. Rank the tournament as you did in that unit and compare your ranking to that in Part c above.

3. Recall from the "Matrix Models" unit how you used powers of an adjacency matrix for a graph to get information about the number of paths of certain lengths in the graph. You also can use powers of an adjacency matrix to find the shortest path length between vertices. Consider the digraph at the right.

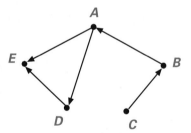

a. Construct an adjacency matrix M for this digraph. List the vertices in alphabetical order. Find M^2, M^3, and M^4.

b. Consider paths from C to E.

- What is the shortest path length from C to E? (Since there are no weights on the edges, the length of a path is the number of edges in the path.)

- Examine the C-E entry in M, M^2, M^3, and M^4. What is the relationship between these entries and the shortest path length from C to E?

c. Explain how you can determine the lengths of shortest paths between vertices in the digraph by examining the powers of the adjacency matrix.

4. In this task, you will investigate Hamiltonian paths and circuits in particular types of graphs.

   a. A **Hamiltonian path** is a path that visits each vertex of a graph exactly once. Thus, a Hamiltonian path is like a Hamiltonian circuit, except the path is not required to start and end at the same vertex. Is it possible for a tree to have a Hamiltonian path? Why or why not?

   b. A **complete graph** is a graph in which every pair of vertices is joined by exactly one edge. Does a complete graph have a Hamiltonian circuit? Illustrate your answer with a complete graph that has five vertices.

   c. A **bipartite graph** is a graph whose vertices can be split into two sets, such that every edge of the graph joins a vertex from one set to a vertex of the other set. A **complete bipartite graph** is a bipartite graph in which every vertex of one set is joined to every vertex in the other set by exactly one edge. For example, here are two complete bipartite graphs:

    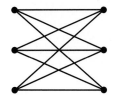

   - Which of these two complete bipartite graphs has a Hamiltonian circuit?
   - Make a conjecture about the kinds of complete bipartite graphs that have Hamiltonian circuits. Defend your conjecture.

Reflecting

1. Why do you think Hamilton's *Traveler's Dodecahedron* and *Icosian* games were not commercially successful?

2. It is estimated that more new mathematics have been developed in the last 20 years than in all the past history of mathematics. In fact, most of the mathematics you have investigated in this unit has been developed in the last few decades, or even more recently. For example, Kruskal's algorithm for minimal spanning trees was developed in the 1950s, and new results related to the Traveling Salesperson Problem are discovered almost every year. Based upon your experience in this unit and additional research, write a short essay on your view of mathematics as a modern, active field.

3. With the rapid development of more and more powerful computers, do you think that any problem eventually can be solved with a brute force method, by having a computer check all possibilities? Or do you think that there is some fundamental limitation to the ability of computers to solve problems? Explain your thinking.

4. The graph models you have studied in the *Contemporary Mathematics in Context* courses include Euler circuits, Hamiltonian circuits, shortest paths, critical paths, minimal spanning trees, vertex coloring, and the Traveling Salesperson Problem.

   a. What kinds of problems do vertex-edge graph models help you to solve?

   b. Which graph models were easiest to learn? To apply?

   c. Of all the vertex-edge graph models you have studied, which is your favorite? Why?

5. The title of this lesson is "Shortest Paths and Circuits," while the title of the unit is "Network Optimization." *Optimize* means to find the best.

   a. Give an example in which the shortest path is not the best path. (Your example does not have to be from this unit.)

   b. A solution to the Traveling Salesperson Problem is a shortest circuit. The brute force method is a guaranteed way to find a shortest circuit. Despite all this, a group of students claims that finding a shortest circuit is not necessarily the best solution to the Traveling Salesperson Problem. Explain what they might mean.

   c. Explain why the best solution method for a small graph may not be the best solution method for a large graph. Give an example.

Extending

1. In Modeling Task 2, page 353, you found Gray Codes by finding Hamiltonian circuits in a graph in which the vertices are 0-1 strings and the edges connect strings that differ in exactly one position.

   a. Redraw the graph you constructed in Modeling Task 2 (for strings of length three) so that it looks like a cube.

   b. It turns out that all Gray Codes can be represented using graphs that look like "cubes" in different dimensions. Draw a graph that can be used to find Gray Codes for strings of length two that looks like a two-dimensional "cube," that is, a square.

   c. For strings of length four, the graph will look like a four-dimensional "cube." Try to draw such a "cube."

   d. Label the vertices of the four-dimensional cube with all the 0-1 strings of length four so two vertices are connected by an edge if the two strings differ in exactly one position. Find a Gray Code by finding a Hamiltonian circuit through the graph.

2. The SHORTCUT software finds shortest paths by using a method called *Dijkstra's algorithm*, named after E. W. Dijkstra, a Dutch mathematician who discovered the algorithm in 1959. Although you must work through the steps of the algorithm very carefully, the basic idea is simple. First, you choose a starting vertex, and you find the vertex closest to the start. Then, you find the next closest vertex to the start, and so on until you have accounted for every vertex. You darken edges and keep track of distances as you go. At the end, you have a shortest path from the start to every other vertex in the graph.

E.W. Dijkstra

Here are the steps of Dijkstra's algorithm:

i. Choose and circle a starting vertex. Since it is the start, write the number 0 next to it, indicating length 0, and circle the number.

ii. Examine *all* edges from the starting vertex. Choose the shortest one (break ties arbitrarily) and darken it. Circle the vertex reached by the shortest edge. Write the length of the edge next to the vertex and circle the number. This number is the length of the shortest path from the start to that vertex.

iii. Examine *all* edges that go from any circled vertex to an uncircled vertex. For each such edge, do the following computation and write the sum on the edge:

$$(\text{length of edge}) + \left(\begin{array}{c} \text{circled number next to circled vertex} \\ \text{out of which the edge comes} \end{array} \right)$$

Darken the edge that yields the smallest sum, and circle the vertex at the end of that edge. Write the computed sum for the edge next to the vertex and circle the number. This number is the shortest distance from the start to that vertex. Erase all the other sums you computed.

iv. Repeat Step iii until there are no more edges from a circled vertex to an uncircled vertex. The circled numbers next to each vertex show the shortest path length from the start to that vertex. The darkened edges show a shortest path from the start to each vertex.

Follow the steps of Dijkstra's algorithm to find a shortest path and the shortest path length from vertex *A* to each of the other vertices in the digraph at the right.

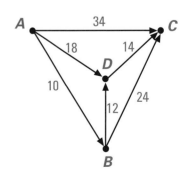

3. The SHORTCUT software uses Dijkstra's algorithm (see Extending Task 2) to find shortest paths. In this task, you will see why Dijkstra's algorithm is sometimes called a "tree-growing" algorithm. Consider the digraph below.

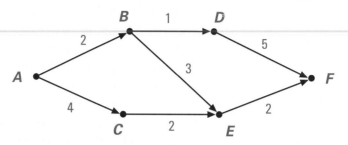

a. Implement Dijkstra's algorithm, either by hand or by using the SHORTCUT program, to find a shortest path and the shortest path length from *A* to each of the other vertices in the digraph.

b. Darken the edges of a shortest path from *A* to each of the other vertices in the graph.

c. Describe the graph consisting of the darkened edges.

■ Is it a tree?

■ Is it a spanning tree?

■ Is it a minimal spanning tree?

4. The nearest-neighbor algorithm you investigated in Activity 9 of Lesson 1 (page 324) did not always produce a minimal spanning tree. A similar algorithm can be applied to the Traveling Salesperson Problem.

a. Look back at the nearest-neighbor algorithm for minimal spanning trees. By slightly modifying that algorithm, write the steps of a nearest-neighbor algorithm for the Traveling Salesperson Problem.

b. Apply your modified algorithm to try to find a solution to the Traveling Salesperson Problem for the weighted graph below.

■ Starting at *A*

■ Starting at *B*

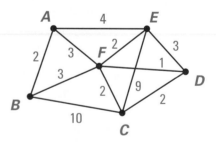

c. What is the shortest possible circuit? Can the shortest possible circuit be obtained by your algorithm using any starting point?

d. Do you think the nearest-neighbor algorithm is a good algorithm for solving the Traveling Salesperson Problem? Explain your reasoning.

5. You have seen in this lesson that there is no known efficient method for solving the Traveling Salesperson Problem. The same is true for finding Hamiltonian circuits and paths. Most experts believe that efficient solutions for these problems will never be found, at least not by using traditional electronic computers. In 1994, computer scientist Leonard M. Adleman of the University of Southern California in Los Angeles opened up the possibility of using nature as the computer to solve these problems. Dr. Adleman successfully carried out a laboratory experiment in which he used DNA to do the computations needed to solve a Hamiltonian path problem. Dr. Adleman stated, "This is the first example, I think, of an actual computation carried out at the molecular level." This method has not been shown to solve all Hamiltonian problems, and the particular problem solved was quite small, but it opens up some amazing possibilities for mathematics and computer science.

a. Below is the graph that Dr. Adleman used in his experiment. All the information in the graph was encoded using strands of DNA, and then the computations needed to find a Hamiltonian path were carried out by biochemical processes. Of course, this graph is small enough that the Hamiltonian path also can be found without gene splicing or conventional computers. Find the Hamiltonian path for this graph.

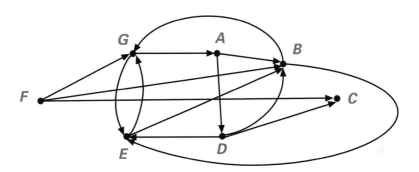

b. Find out more about this groundbreaking experiment in "molecular computation" by reading some of the articles below. Write a short report incorporating your findings.

- Adleman, Leonard M. Molecular Computation of Solutions to Combinatorial Problems. *Science*, November 11, 1994.

- Delvin, Keith. Test Tube Computing with DNA. *Math Horizons*, April, 1995.

- Kolata, Gina. Scientist At Work: Leonard Adleman: Hitting the High Spots of Computer Theory. *The New York Times*, Late Edition, December 15, 1994.

Fan Chung

6. Vertex-edge graphs are part of an area of mathematics called discrete mathematics. A leading contemporary researcher in discrete mathematics is Fan Chung, a mathematician who earned her doctorate in 1974 and has worked at Bell Labs, Harvard University, and the University of Pennsylvania. Although many of the problems that Dr. Chung works on look like games, often they have important applications in areas like communication networks and design of computer hardware and software.

Consider one such problem. In the graph below, suppose a person is standing at each vertex. The first letter of each vertex label is the name of the vertex; the second letter is the destination that each person must reach by walking along edges of the graph. The goal is for all the people to walk to their destinations without overusing any edge. By trying to solve a problem like this, you can find out how accessible a network is—that is, if it has any "bottlenecks" where there is excessive traffic.

a. Assume that an edge is "overused" if it is used more than twice. Find routes for all walkers so that everyone reaches their destination but no edge is overused. Are other routes possible?

b. Describe the strategy you used to find the routes.

c. State, and try to solve, at least one other problem related to this situation.

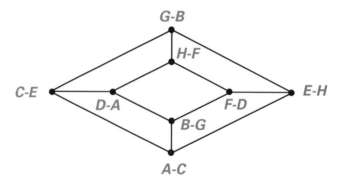

Looking Back

In this unit, you have investigated the Traveling Salesperson Problem and several new vertex-edge graph models: trees, minimal spanning trees, shortest paths, and Hamiltonian circuits. In order to effectively use these models to solve problems, you need to know what they are, when to use them, and how to use them. In this lesson, you will review and apply these models in new contexts.

1. Consider the map below of a region of Kentucky.

For each of the following questions, state which graph model could be used to model the situation. You do not need to answer the questions, just state which graph model applies. Explain your choice of models.

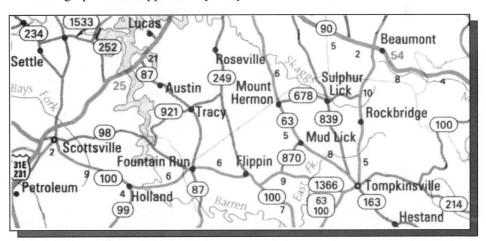

Map © 1997 by Rand McNally, R.L. 97-S-79

a. Suppose a truck driver must visit each town in the region to deliver packages, beginning and ending at Tompkinsville. What is the minimum number of miles the driver would have to drive?

b. What is the shortest route from Scottsville to Rockbridge?

c. Since snowstorms are uncommon in Kentucky, the transportation departments do not have a lot of snowplows standing by. So when it does snow, they need to have an efficient plan for plowing the roads. Suppose the goal is to plow just enough roads so that there is a way to get from every town to every other town on plowed roads. What is the minimum number of miles that must be plowed in order to achieve this goal?

2. The influence one person exerts over another is an important question that sociologists study. Consider a group of five associates in an architectural firm.

A sociologist asks each person to fill out a questionnaire identifying the one person in the group whose opinion the respondent values the most. If person X values person Y's opinion, then it is assumed that person X is influenced by person Y. The sociologist wants to find out which person has the most influence on the group. The results of the questionnaire are tabulated below.

Results of Influence Study

Member of the Group	Person Whose Opinion Is Valued Most
A	D
B	A
C	B
D	B
E	D

Use these data and some kind of network optimization to determine who is the most influential person in the group. Justify your conclusion.

3. You have used vertex-edge graphs to model a variety of situations. You also can use graphs to represent and analyze relationships among the new concepts that you are learning. This is done using a type of graph called a *concept map*. In a concept map, the vertices represent ideas or concepts, and edges illustrate how the concepts are connected. The edges may or may not have labels. The first step in building a concept map for some area of study is to list all the concepts you can think of. Here is the beginning of such a list for this unit on network optimization:

- minimal spanning tree
- Traveling Salesperson Problem
- best-edge algorithm

a. Add to this list. Include all the concepts from this unit that you can recall.

The next step in building a concept map is to let the concepts in the list be the vertices of a graph, and then draw edges between vertices to show connections between concepts. The edges should be labeled to show *how* the vertex concepts are connected. There are many different concept maps that can be drawn. The beginning of one concept map for this unit is drawn below.

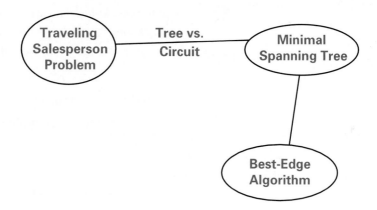

b. Interpret the sample concept map above.

- Why is there an edge from *best-edge algorithm* to *minimal spanning tree*?
- Why shouldn't an edge be drawn from *best-edge algorithm* to *Traveling Salesperson Problem*?
- The edge between *Traveling Salesperson Problem* and *minimal spanning tree* is labeled *tree vs. circuit*. Explain why a solution to the Traveling Salesperson Problem can be described as a minimal spanning circuit, and then explain why the edge is labeled *tree vs. circuit*.

c. Complete the concept map by adding all the concepts (vertices) that you listed in Part a along with appropriate edges that show connections among the concepts.

Checkpoint

In this unit, you have studied important concepts and methods related to network optimization.

a Compare your concept map from Task 3 with those of other students. Discuss similarities and differences.

b Optimization or "finding the best" is an important theme throughout this unit. Describe three problem situations from the unit in which you "found the best." In each case, explain how you used a graph model to solve the problem.

c Describe one similarity between minimal spanning trees and shortest paths. Describe one difference.

d Describe one similarity between minimal spanning tree problems and the Traveling Salesperson Problem. Describe one difference.

e Describe one similarity between the Traveling Salesperson Problem and shortest path problems. Describe one difference.

f In this unit, you explored a variety of algorithms and methods for solving network optimization problems, including best-edge algorithms and brute force methods. For each of these two solution procedures, do the following.

- Describe the basic strategy.
- Give some examples of problems that can be solved using the procedure.
- List some advantages and disadvantages of the procedure.

Be prepared to share your examples and descriptions with the entire class.

On Your Own

Write, in outline form, a summary of the important mathematical concepts and methods developed in this unit. Organize your summary so that it can be used as a quick reference in future units and courses.

Geometric Form and Its Function

Flexible Quadrilaterals

The geometric form of an object often is influenced by its function, whether the object is naturally occurring or engineered. Because bee honeycombs are composed of hexagonal prisms, the individual cells not only hold honey, but they also fit together with no gaps. Thus, the particular geometric form serves two functions. The designs of objects by architects and engineers are similarly influenced by the objects' functions. For example, buildings and bridges must be rigid. Motors and bicycles involve circular motion. The suspension system of an automobile helps control motion of the vehicle.

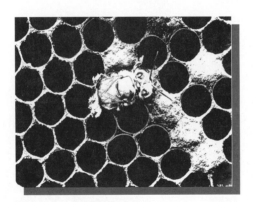

Triangles, quadrilaterals, and circles are geometric shapes commonly used by designers. A triangle is rigid; a circle turns easily about its center; and the appearance of a quadrilateral can be changed by changing the measures of its interior angles, without changing the length of any side.

Shown below is a simple linkage made from plastic strips joined with brass paper fasteners. Imagine the various ways you could change the shape of this linkage. The diagrams below show two possible shapes for the linkage.

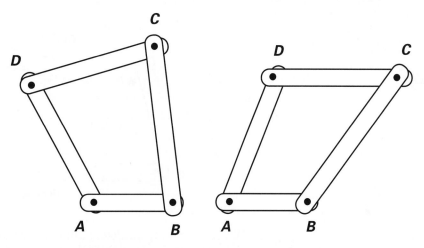

Think About This Situation

Consider the following geometric and design questions related to a quadrilateral linkage.

a How could you change a linkage with the shape on the left above so that it has the shape on the right?

b If a model of a quadrilateral has hinged vertices, what features of the model can change? What features cannot change?

c Imagine holding strip *AB* still and turning strip *CB* clockwise about point *B*. What will happen to strips *CD* and *AD*?

d Imagine placing the linkage on a sheet of paper, inserting a pencil through point *B*, and holding the linkage firmly at point *A* as you move the pencil. What shapes could you draw? What shapes would you draw with another pencil inserted through point *D*, if you moved the pencil in point *B*?

e What could you do to the linkage shown above to make it rigid?

In this unit, you will explore some of the ways in which triangular, quadrilateral, and circular shapes are used to perform everyday functions and help solve problems. You will become more aware of how function and geometric form are related.

INVESTIGATION 1 Using Quadrilaterals in Linkages

Quadrilaterals are rigid when *triangulated*, but when they have no diagonal, they change shape readily. Mechanical engineers use the nonrigidity of quadrilaterals to their advantage in the design of linkages. A **quadrilateral**, or **4-bar**, **linkage** like the one shown below can change shape by pivoting about its vertices. At the same time, the linkage retains some of the characteristics of the original shape.

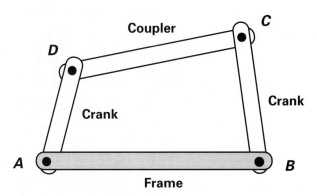

When 4-bar, or quadrilateral, linkages are used in mechanical contexts, one of the four sides is fixed so it does not move. It is called the *frame*. The two sides attached to the frame are called *cranks*; one is called the *driver* and the other is called the *follower*. The driver is the crank most directly affected by the user of the linkage. The cranks are connected by the *coupler*.

1. These features of a quadrilateral linkage can be seen in handcars like the one shown below. Make a cardboard model of this handcar. Use paper fasteners to put it together.

   a. What are the cranks? Where are these shown in the diagram? Describe their motions.

   b. What are the endpoints of the coupler?

   c. What points determine the frame?

   d. Does the driver crank cause the follower crank to rotate?

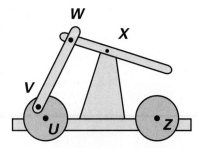

In 1883, a mathematician named Grashof suggested the following general principle for quadrilateral linkages:

> *If the sum of the lengths of the shortest and longest sides is less than or equal to the sum of the lengths of the remaining two sides, then the shortest side can rotate completely.*

2. Obtain some plastic or cardboard strips and paper fasteners. Working with a partner, make a model to investigate the truth of Grashof's principle for the situation described on the following page.

a. Design your quadrilateral linkage *ABCD* so that strip *AB* is the longest side, strip *AD* is the shortest side, and *AB* + *AD* < *BC* + *CD*. What are the lengths of the sides of your model?

b. Does strip *AD* rotate completely?

c. Describe the motion of strip *BC*. Through what part of a circle does it move?

d. How is your model of the handcar in Activity 1 related to this linkage?

e. How could this linkage be used to make a stirring mechanism? To make a windshield wiper work?

f. Compare your responses with those of another pair of students.

3. Next, work with a partner to make a model of a *parallelogram linkage* using plastic or cardboard strips and paper fasteners to connect the sides.

a. Explain how you know your model is a parallelogram.

b. Choose a side to be the frame and hold it fixed on your desk as the base of the quadrilateral. Move one of the vertices on the opposite side (the coupler) to the right and then to the left. As the vertex moves, the shape of the linkage changes. Describe the changes you see. What properties of the shape, if any, do *not* change?

c. Move the coupler back and forth in small increments. With each move, mark the location of each vertex of the coupler. What patterns do these points follow?

d. Write an explanation that will convince other members of your group that the patterns you observed in Part c describe the actual paths the vertices follow.

e. Choose a point (but not a vertex) on the side opposite the frame. Move that side. What path does the chosen point follow? Explain why the point moves along the path it does.

4. The diagram below depicts a windshield wiper mechanism as found on buses and tractor-trailers. The wiper blade is attached to the mechanism in a fixed position.

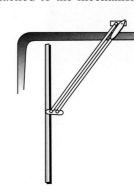

a. Make a sketch of this mechanism. Label the frame, cranks, and coupler.

b. Explain why this is a parallelogram linkage.

c. As the linkage moves, what paths do the ends of the wiper blade follow?

d. If the wiper blade is vertical (as shown) when the mechanism is at the beginning of a cycle, describe the positions of the blade when the mechanism is one quarter of the way through its cycle and when the mechanism is halfway through its cycle.

e. Sketch the region of the windshield that the blade keeps clean when in use.

Checkpoint

Summarize key ideas about quadrilateral linkages from your work in this investigation.

ⓐ Describe a quadrilateral linkage in which both the driver and follower cranks may make complete revolutions.

ⓑ List as many characteristics of a parallelogram linkage as you can. Which of these characteristics make it useful in mechanical devices?

Be prepared to share your descriptions and thinking with the entire class.

▶ On Your Own

Understanding the body mechanics involved in various physical activities is important to sports physicians and trainers. The diagram below shows a person pedaling a bicycle. Key points in the pedaling motion are labeled.

a. What kind of linkage is represented by *ABCD*?

b. Identify the frame, the coupler, the driver crank, and the follower crank.

c. What modifications to the situation would allow it to be modeled by a parallelogram linkage? Should a sports trainer recommend these modifications? Explain your reasoning.

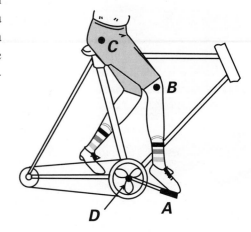

INVESTIGATION 2 Linkages and Similarity

Some quadrilateral linkages can change "back-and-forth" motion into rotary motion and vice versa. In a parallelogram linkage, both cranks can rotate completely. In addition to being used in mechanical devices, the parallelogram linkage serves as the basis for a linkage called a *pantograph*. The following pantograph is held firm at point *F*; the rods are hinged at *B*, *D*, *E*, and *G* so that *BDGE* is always a parallelogram. The other dots are holes that can be used to change the size of parallelogram *BDGE*.

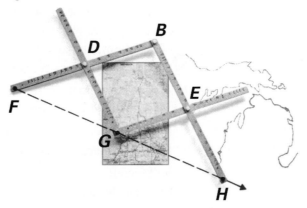

A pantograph can be used to enlarge or reduce the size of a map, a picture, or other figure while retaining its shape. In the following activities, you will explore how a pantograph works and why a figure and its pantograph image are *similar*.

1. Working with a partner, construct a model of a pantograph like the one shown below. Connect four linkage strips at pivot points *B*, *D*, *G*, and *E* with paper fasteners.

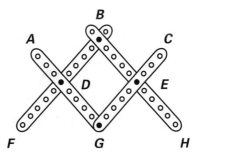

   a. What kind of parallelogram is *BDGE*? Explain your reasoning.

   b. Experiment to see if you can figure how to use a pantograph.

   c. What changes and what remains the same for quadrilateral *BDGE* as the pantograph is used?

   d. Compare your responses to Parts a and c with the responses of other group members.

2. Using a copy of the pantograph drawing in Activity 1, with points *B*, *D*, *E*, *F*, *G*, and *H* labeled, complete the activities below.

   a. Outline △*FDG*. Then use a different colored pen or pencil to outline △*FBH*.

   b. Compare △*FDG* and △*FBH*. How are their sides related? Their angles? How are the two triangles related?

3. Again working with your partner, use your model to explore the following questions. Compare your findings with those of other members of your group.

   a. Draw a line across a sheet of paper and fix point *F* near one end. Place point *G* somewhere else on the line. Where is point *H* in relation to the line? If point *G* moves along the line, what path does point *H* follow? From your observations, how do you think points *F*, *G*, and *H* are related? Is this relation always true for other locations of points *F*, *G*, and *H*?

   b. Adjust the pantograph so that *FG* is 6 centimeters. How long is $\overline{FH}$? How does length *FH* compare to length *FG*?

   c. If *FG* is 9 centimeters, predict the length of $\overline{FH}$. Check your prediction using the pantograph.

   d. In general, for this setup of the pantograph, what will be the relation between the lengths *FG* and *FH*? Write this relation in equation form: *FH* = _____ .

4. Next, mark a point *F* on a large sheet of paper. Then position and hold your pantograph on the paper so that the points *F* on the pantograph and on the paper are aligned. Using the drawing below as a guide, mark points *P* and *Q* on your paper. Insert a pencil through the hole at point *H*.

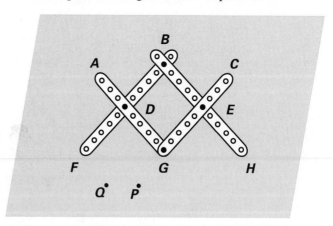

   a. Find the image of *P* by placing *G* over *P* and marking the location of *H*. Label your mark as the image point *P′*. What is true about *F*, *P*, and *P′*?

   b. Find the image of *Q* in the same manner. Label it *Q′*. What is true about *F*, *Q*, and *Q′*?

   c. Compare the distances *PQ* and *P′Q′*. How does this relationship compare with your observations in Part b of Activity 2, which focuses on the corresponding sides of △*FDG* and △*FBH*?

d. Choose two new locations for points *P* and *Q* and repeat the process described in Parts a–c.

e. In general, how is the distance between any two points *P* and *Q* related to the distance between the image points *P′* and *Q′*?

- Explain why this will always be the case.
- Explain how this relationship can be seen in the relationship between corresponding sides of △*FDG* and △*FBH*.

5. Now reassemble your pantograph as illustrated in the diagram below.

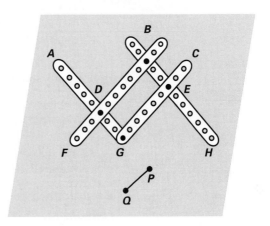

a. Compare △*FDG* and △*FBH*. How are their sides related? Their angles? How are the two triangles related? What is the scale factor?

b. Draw a segment, $\overline{PQ}$, that is 5 cm long, and predict the length of the image $\overline{P′Q′}$. Explain the basis for your prediction and then check it using the pantograph.

c. How will the pantograph affect distances if it is assembled so that $FB = k \cdot FD$?

6. On a large sheet of paper, draw a right triangle with legs 5 and 8 centimeters. Position the triangle so that there is room to create an enlargement of it using your pantograph.

a. Using your pantograph as assembled in Activity 5, find the image of the right triangle. Is it sufficient to find the images of the three vertices and then connect them to get the image triangle? Explain your reasoning.

b. How are the corresponding sides of the original triangle and the image triangle related? How are the two triangles related?

c. What is the area of the original triangle? The image triangle? Describe the relationship between these areas.

d. Reset your pantograph so that it multiplies distances by 2. Use it to find another image of the original right triangle you drew.

e. How are corresponding sides of the original and image triangles related? How are the areas of the two triangles related?

f. Explain how the size of an angle in a figure and the size of the corresponding angle in its pantograph image are related.

7. Consider how you would assemble your pantograph so that it multiplies distances by 4.

    a. Draw a triangle with area 10 cm^2. Predict the area of the image triangle made with this pantograph.

    b. Compare your predictions with those of another group. Resolve any differences.

Checkpoint

Suppose a pantograph is to be used to make an enlargement of a shape using a scale factor of k.

a How would you assemble the pantograph?

b How will the two shapes be related?

c How will corresponding lengths found in the two shapes be related? How will measures of corresponding angles be related? How will areas of corresponding regions be related?

Be prepared to share your responses and explain your reasoning.

On Your Own

Suppose a pantograph is assembled to enlarge this cartoon by a scale factor of 6. Assume points on the pantograph are labeled as in Activity 1.

a. What are the lengths of the sides of $\triangle FDG$? Of $\triangle FBH$?

b. How are the angles of $\triangle FDG$ and $\triangle FBH$ related?

c. Compare the areas of $\triangle FDG$ and $\triangle FBH$.

d. What will be true about the shape and size of the pyramid in the enlarged image of the cartoon?

"We had a little problem with the decimal point."

©1995; Reprinted courtesy of Bunny Hoest and Parade Magazine.

Modeling

1. A common adjustable desk lamp is shown in the diagram below. The pivots at the labeled points are snug, but they will allow pivoting to adjust the lamp position.

   a. Identify the parallelogram linkages that are used in this lamp.

 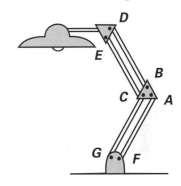

   b. Describe how the position of the light shade changes as you make *AFGC* vertical.

   c. Keeping *AFGC* in one position, how does changing the position of *BDEC* change the position of the light shade?

   d. Make a cardboard model of this lamp linkage. Investigate the angle the light shade makes with the horizontal for various positions of the parallelogram linkages.

2. Trays in sewing boxes, tool boxes, and tackle boxes often use a linkage system to make accessible two or more trays by lifting only the top one. When this is done, the linkage allows them to close and fit nicely inside the box. Examine the two boxes shown below.

   a. For each box, do the trays remain horizontal when opening them? When they are completely opened? Explain your reasoning.

b. Make a sketch of a two-dimensional side view of each box showing how the linkage works.

c. Sketch a side view of a linkage system for a four-tray box.

3. Sketch a pantograph assembled from strips that are 30 cm from the center of one endhole to the center of the other endhole. Additional holes are punched, each 1 cm from its neighbors. Label the pantograph in the same way as the pantograph on page 373.

a. What whole-number scale factors are designed into this pantograph? Describe the lengths of $\overline{FD}$ and $\overline{FB}$ in each case.

b. How would you set up the pantograph to get a scale factor of $\frac{3}{2}$?

c. How would you set up the pantograph to get a scale factor of 2.5?

d. Suppose you have two shapes, and Shape II is the image of Shape I using the pantograph in Part c.

- If a segment in Shape I measures 3 cm, what is the measure of the corresponding segment in Shape II?

- If a segment in Shape II measures 9 cm, what is the measure of the corresponding segment in Shape I?

- If the area of Shape I is 80 cm², what is the area of Shape II?

- If the area of Shape II is 100 cm², what is the area of Shape I?

Compass Productions, Long Beach, CA

4. A common method of designing a pop-up page or greeting card is based on a parallelogram. Tabs attached to facing pages form two consecutive faces of a prism with a parallelogram base. The other two faces are attached to the tabs and serve as props for the picture. As a page or card is turned, the prism unfolds and the picture pops up.

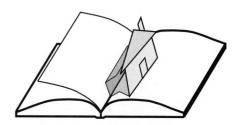

The dark gray shaded area indicates the pop-up tab that pulls the picture off the page as the page is turned.

a. Use this method to construct a paper model of a pop-up picture. Use a piece of paper folded in the middle as consecutive pages of a book. A second piece of paper is needed to carry the picture. Make your model carefully. Test it. Modify it until it works well.

b. Why does the picture in your model lie flat when the book is closed?

c. Why does the picture in your model lie flat when the book is wide open?

d. When will the picture be positioned perpendicular to the right-hand page?

e. When will the picture make a 120° angle with the right-hand page?

f. Could the parallelogram base of the prism be replaced by any other quadrilateral? If so, which ones? If not, explain your reasoning.

5. A windshield wiper on an ordinary automobile oscillates back and forth to remove water.

   a. Make a sketch of a linkage that could be used to perform this function.

   b. In your sketch, identify the frame, cranks (driver and follower), and coupler.

   c. How does this type of mechanism differ from the wiper mechanism for a tractor-trailer, described in Activity 4 of Investigation 1 (page 371)?

   d. Suppose a 36-cm wiper blade is attached at its midpoint to the end of a 36-cm wiper arm and oscillates through a 100° angle. Find the area of the windshield swept clean by this wiper.

Organizing

1. Consider a pantograph set with a scale factor of 3 as shown. Imagine point *F* fixed at the origin of a coordinate system.

   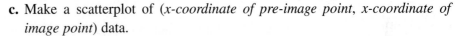

   a. If you place *G* over the point (2, 3) on the coordinate system, what are the coordinates of the corresponding point at *H*?

   b. Answer the question asked in Part a for several additional points as well. What pattern do you see relating the coordinates of the points at *G* and their images at *H*?

   c. Make a scatterplot of (*x-coordinate of pre-image point*, *x-coordinate of image point*) data.

   d. Find an algebraic model that best summarizes the pattern in the scatterplot. Use your model to predict the *x*-coordinate of an image point whose pre-image has an *x*-coordinate of 18.

2. Consider a pantograph assembled with a scale factor of 3 as in Organizing Task 1. Imagine point *G* fixed at the origin of a coordinate system.

   a. If you place *F* over the point (3, 6) on the coordinate system, what are the coordinates of the corresponding point at *H*?

   b. Consider placing *F* over several additional points in all four quadrants and find their images as you did in Part a. What pattern do you see relating the coordinates of the points and their images?

   c. Make a scatterplot of (*y-coordinate of pre-image point*, *y-coordinate of image point*) data. What kind of relation do you observe?

   d. Write an equation describing the relation you observed in Part c. Use your equation to predict the *y*-coordinate of an image point whose pre-image has a *y*-coordinate of –24.

   e. What is the scale factor in using a pantograph in this way?

3. In the "Patterns of Location, Shape, and Size" unit, you investigated size transformations in a coordinate plane. Recall that a size transformation with magnitude 3 was described by the following rule:

pre-image   image

$$(x, y) \quad \rightarrow \quad (3x, 3y)$$

   a. On a coordinate grid, graph the following points and their images under the size transformation above.

   $A(1, 3)$ $B(2, -3)$ $C(-2, -1)$ $D(-2, 4)$

   b. How is quadrilateral $ABCD$ related to its size transformation image?

   c. How could you achieve the same enlargement of quadrilateral $ABCD$ using a pantograph?

4. Think about finding the image of a shape using a pantograph with scale factor 2, and then finding the image of that image with a scale factor 3 pantograph.

   a. How are the original shape and the final image related with respect to lengths? With respect to areas?

   b. If you wished to accomplish this enlargement using a single pantograph, what scale factor would be needed? Explain your reasoning.

   c. Suppose you used a combination of pantographs with scale factors h and k. How should the scale factor of a single pantograph be set to accomplish the same magnification?

Reflecting

1. What did you find most challenging in your work with quadrilateral and parallelogram linkages and pantographs? What suggestions would you offer to others who are just beginning this lesson?

2. Talk with someone who enjoys model railroads, or visit a local hobby shop in preparation for answering the following questions.

   a. How is a parallelogram linkage used on models of old train locomotives?

   b. How is this use different from the other uses of linkages with which you are familiar?

3. Explain how you can determine the scale factor of a pantograph using a diagram similar to those found in Investigation 2 (pages 373–375).

4. Find an example of a parallelogram linkage different from the examples in this lesson. Write a paragraph describing its purpose and how it works.

Extending

1. Make the linkage shown below, in which $AF = BE = CD$, $BC = ED = 2AB$, and $AB = FE$. Note that there is only one strip (shown horizontally) connecting points D, E, and F. Similarly, there is only one connecting A, B, and C.

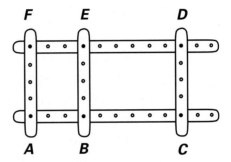

a. Hold strip EB fixed and move strip AF. Describe the motion of strip CD.

b. Hold strip AF fixed and move strip BE. Describe how strip CD moves compared to strip BE.

c. When strip AF is fixed and strip BE moves, what can you say about the movement of points on the linkage that lie between points B and C? Between points A and B? Between points D and C?

d. Fix a strip between points F and B and move strip CD. What happens? Explain why this occurs.

2. Expandable safety gates, such as the one shown in the diagram below, are based on parallelogram linkages. Pivots are placed at each intersection point so that the gate will close and open.

a. Place a sheet of paper over the diagram and then trace out linkages that would have scale factors of 3, 4, and 1.5 if they were pantographs.

b. The gate network can be used to design linkages like a pantograph that will enlarge (or reduce) shapes. For example, the two-rhombus linkage below, with point *F* fixed, enlarges a shape at *G* to a shape at *H*, with scale factor 3. Model this linkage to check that it enlarges with scale factor 3.

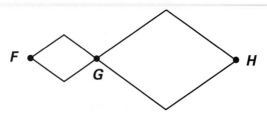

c. Draw two additional linkages based on the safety gate that have a scale factor of 3. How must the three points *F*, *G*, and *H* be related in each case?

3. Connect three strips *DA*, *AB*, and *BC* (with *DA* = *BC*) with paper fasteners at *A* and *B*. Fasten ends *C* and *D* to a card so that *DC* = *AB*. Hold the card perpendicular to the floor and adjust the fasteners to allow strip *AB* to swing easily back and forth.

a. Hold $\overline{CD}$ horizontal and swing strip *AB* back and forth. Does strip *AB* remain horizontal as it swings? Why or why not?

b. What path does a point on strip *AD* or strip *BC* follow as strip *AB* swings back and forth?

c. Investigate the paths of points *A*, *B*, and additional points on strip *AB* as the strip swings back and forth. (Marking positions on paper will help.)

d. Explain how this linkage could be used to design a child's toy.

4. The linkage shown below is made up of three rhombuses. The longer bars such as *PQ* are three times the lengths of the shorter bars such as *XP*.

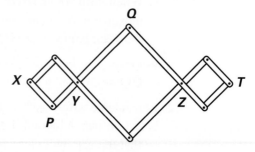

a. Suppose you hold the linkage fixed at *X* and copy a shape at *Y* with pencils inserted at both *Z* and *T*. What is the scale factor relating the shape copied at *Y* to the shape copied at *Z*? To the shape copied at *T*?

b. Suppose you fix *X* and copy a shape at *Z* with pencils at both *Y* and *T*. What is the scale factor relating the shape copied at *Z* to the shape copied at *Y*? To the shape copied at *T*?

c. How can the linkage be used to produce similar shapes with each scale factor below?

■ $\frac{1}{4}$ ■ $\frac{3}{4}$ ■ $\frac{3}{2}$ ■ $\frac{2}{3}$

d. Make a model of the linkage to check your conclusions in Parts a–c.

e. Design three different linkages that are capable of making a copy similar to the original with scale factor 4.

5. Assemble a pantograph with scale factor 2, labeled as in Investigation 2.

a. If you fasten the pantograph at vertex F, trace a shape using vertex H, and draw the image with a pencil inserted through the hole at G, how are the two shapes related?

b. What is the scale factor for this arrangement?

c. Set up your pantograph so that a reduction by a scale factor of $\frac{1}{3}$ can be accomplished. What point is fixed? Where do you place the shape to be copied?

d. How can you check that your model in Part c is properly assembled?

e. Companies that make very small letter etchings use reducing pantographs. In terms of our pantograph labeling, they have model letters that are traced at H while the etched letters are produced at G. What are some reasons such a procedure might be used rather than etching the small letters directly?

f. How might the etching company control the variation in its etchings?

Triangles and Trigonometric Ratios

Quadrilaterals, especially parallelograms, have many uses because they can pivot about their vertices to change shape without changing the length of any sides. Triangular shapes, whether their vertices can pivot or not, are rigid once the lengths of the sides are specified.

The rigidity of triangles is often used to make more complex structures rigid. In contrast, the truck crane shown below operates by adjusting the length of a side of a triangular structure. As the hydraulic cylinder lengthens, the boom rises.

Boom
Hydraulic Cylinder

Think About This Situation

Examine the design of the truck crane shown above.

a About what point does the boom pivot?

b Why is the boom structure rigid?

c Under what conditions could the boom be raised so that it is perpendicular to the truck bed? What kind of triangle would be formed?

d Describe other situations in which adjusting the length of one side of a triangular shape serves a useful function.

INVESTIGATION 1 Triangles with a Variable-Length Side

In the first investigation of this lesson, you will explore uses of adjustable triangular shapes, with a side that can vary in length.

1. Make a model of a triangle with a variable-length side as illustrated below. Make strip *AB* 10 cm long and strip *BC* 16 cm long from endhole to endhole.

   a. What is the maximum length needed for strip *AD*? Make strip *AD* with holes 2 cm apart (or draw a segment, $\overline{AD}$, on your paper and mark points 2 cm apart).

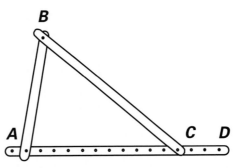

   b. Use strip *AD* to form △*ABC* with a variable side *AC*. When you change the length *AC*, what else changes?

   c. Adjust the length of $\overline{AC}$ in equal step sizes. At each step, use a ruler and protractor to obtain and record the measurements indicated in the table below. Share the workload among members of your group.

Length *AC*	Measure of ∠*A*	Measure of ∠*B*	Measure of ∠*C*	Perpendicular Distance From Point *B* to $\overline{AD}$
___	___	___	___	___
___	___	___	___	___

   d. Using the lengths of $\overline{AC}$ on the horizontal axis, make individual scatterplots of the following data pairs. Describe any patterns you see.
   - (*length AC, measure of ∠A*)
   - (*length AC, measure of ∠B*)
   - (*length AC, measure of ∠C*)
   - (*length AC, perpendicular distance from point B to $\overline{AD}$*)

   e. Describe how each of the other variables changes as length *AC* changes in equal amounts.

   f. Compare the scatterplot of (*length AC, measure of ∠A*) with that of (*length AC, measure of ∠B*).

   g. Compare the scatterplot of (*length AC, measure of ∠C*) with that of (*length AC, perpendicular distance from point B to $\overline{AD}$*).

As one side of a triangle changes, the height and all the angles of the triangle change. These corresponding changes are used in the world around you to serve various functions. Consider first how properties of a variable-sided triangle are used in the design of lawn chairs.

2. Side views of two reclining lawn chairs are shown.

   a. Examine the first design at the right.

   ▪ How many seating positions does this chair have?

   ▪ Identify the three vertices of a variable-sided triangle.

   ▪ Explain how this chair works in terms of a variable-sided triangle.

   b. Here is another way to design a reclining chair. The arms can hook the legs at point *C*. The front leg is attached to the seat at *F*; the back leg hits a rod at the back of the seat (*E*), which keeps that leg from collapsing.

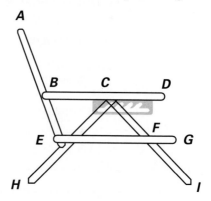

   ▪ Identify the three vertices of a variable-sided triangle.

   ▪ How many positions does this chair have?

   c. Make a sketch or a two-dimensional side-view model of a variable-sided triangle like the one used in the chair in Part b. In your model or with your sketch, find the maximum reclining angle. Find the smallest reclining angle.

In Activity 2, the angle of recline of the chairs varied as the length of a variable side of a triangle changed. In other applications of a triangle *ABC* with variable side *AC*, the height of point *B* above side *AC* is important, rather than the size of an angle.

3. An automobile jack that comes with some makes of automobiles is shown below. Points *A* and *C* are connected by a long threaded rod.

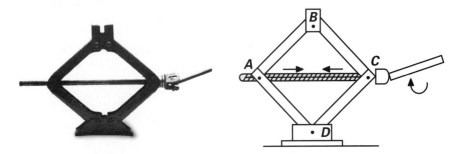

a. Model this jack using plastic or cardboard strips. You can simulate the threaded rod by cutting a slit in a strip of cardboard and using that strip for the rod *AC*. Washers at *A* and *C* may make the model work more smoothly.

b. Examine your model. Is this linkage rigid? Explain why or why not.

c. Identify the triangles that have a side that may be adjusted in length.

d. How could this linkage be stored in a car? About how long would the threaded rod need to be?

e. As the threaded rod connecting points *A* and *C* is turned at a constant speed, *A* and *C* approach each other at a constant rate. What path does point *B* follow if the base of the jack *D* is in a fixed position?

f. Approximately how high is point *B* when the model jack is fully extended?

4. Another kind of automobile jack, shown in the following diagram, is used on racetracks and in garages that do quick tire changes. The jack has wheels to move it around. Bar *BD* has one end that can be moved toward and away from point *A* by turning a threaded rod. In the diagram, the bar ends at point *C*, but it can be moved as far as point *E*.

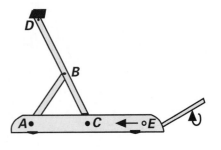

a. Make a model of this jack using plastic strips or stiff cardboard. The length of line segment *AE* should be 20 cm. The lengths *AB*, *BC*, and *BD* should be 10 cm each. Fix point *A* and let point *C* slide along a slot *AE*. Experiment with your model to see how it works.

b. As point *C* moves toward point *A*, what path does point *D* follow? Why is your observation important to the safe use of the automobile jack?

c. As point C moves toward point A, what path does point B follow?

d. Choose a point between points D and B and trace its path as point C moves toward point A.

e. Collect data relating the distance AC to the distance AD for different positions of point C. Describe the rate of change in AD as AC decreases in unit increments.

5. Now adjust your model from Activity 4 so that the measure of $\angle C$ is $45°$.

    a. Find the height of point D.

    b. What is the length of base AC?

    c. What kind of triangle is $\triangle DAC$?

Checkpoint

Triangular shapes often are useful because they are rigid. But designing triangular shapes to permit one side to vary in length often makes them even more useful.

a Describe the characteristics of a variable-sided triangle that make it useful in the design of everyday objects.

b Suppose $\triangle ABC$ has an adjustable side AC. How would the measure of $\angle C$ and the height of $\triangle ABC$ (from vertex B to side AC) be affected as the adjustable side changes in length?

Be prepared to share your conclusions and reasoning with the class.

On Your Own

Architects sometimes use large tables that can be tilted forward, as shown in the photo at the right. How could you use a variable-sided triangle to make such a table that has three tilt positions? In your design, what is the size of the angle formed by the stand and the variable side when the tabletop is horizontal? When the top is at a $45°$ angle to the horizontal?

Modeling

1. A cold frame is a box used to grow young plants in the spring. Traditionally, the top of the box is made of glass to let in light. The top can be propped open so the plants become accustomed to actual weather conditions before they are transplanted outside the box. One cold frame has a top measuring 120 cm by 80 cm, hinged along the 120-cm edge. It can be opened to a 10°, 20°, 30°, or 40° angle.

   a. Design a system that will accomplish the desired openings using only one bar to prop open the top.

   b. Describe the measurements needed in your propping system.

   c. If you wanted to add to your design a 25° opening, where would the prop be placed?

   d. How high above the horizontal frame is the front of the lid when the angle is 20°?

2. Most ironing boards can be adjusted to different heights. One ironing board, designed similar to the one shown here, has legs that are each 110 cm. Possible working heights are 90 cm, 85 cm, 80 cm, 75 cm, and 70 cm.

   a. Make a scale model of the board described above.

   b. Describe how the legs should be attached to the board.

   c. How is a variable-sided triangle used in the design?

   d. For each working height, give the side and angle measurements of the adjustable triangles.

3. A carnival ride consists of six small airplanes attached to a vertical pole. (See the diagram below.) As the pole rotates, the planes fly around the pole. A rider can control the height of the plane by changing the length of the hydraulic cylinder attached at point *C*. In a typical design, *BD* = 4 m, *BA* = 1.5 m, *BC* = 1.5 m, *BE* = 1.5 m, and *AC* varies between 1.5 and 2.2 m.

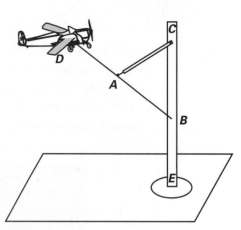

a. Make a model of this carnival ride.

b. What is the smallest measure of ∠*ABC*? How far above the ground is the plane for that smallest angle?

c. If the hydraulic cylinder is fully extended, what is the measure of ∠*ABC*?

d. What variation does the height of the airplane have as ∠*ABC* changes from its minimum to its maximum size?

e. How should the hydraulic cylinder be adjusted so that the plane will fly 2.5 meters above the ground?

Organizing

1. If necessary, re-enter the (*length AC, measure of ∠A*) and (*length AC, measure of ∠B*) data you collected for Activity 1 of Investigation 1 (page 385) into the data lists of your graphing calculator or computer software.

a. Compare the scatterplots of (*length AC, measure of ∠A*) and (*length AC, measure of ∠B*).

b. For each plot, investigate modeling the pattern in the data with different types of algebraic rules: linear, exponential, and power. Are any of these models a good fit for the data? Explain the reasons for your conclusions.

   c. Examine the scatterplots of other pairs of the measures you collected for Activity 1 of Investigation 1. Which of these plots are modeled well by a line? Find an equation of each such line and explain what the slope and *y*-intercept of the model mean.

   d. What can you say about the other data patterns in Part c?

2. Make a linkage in the shape of a rhombus.

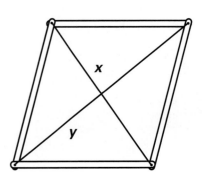

   a. Experiment by changing the rhombus linkage from short (vertically) to tall. Using small changes, collect data on the lengths *x* and *y* of the diagonals.

   b. Make a scatterplot of the (*x*, *y*) data. Describe any patterns you see.

   c. Analyze the pattern of change.

     ■ Does the pattern of change appear to be linear? Explain.

     ■ Does the pattern of change appear to be exponential? Explain.

     ■ Does the pattern of change appear to be that of a power or a quadratic rule? Explain.

3. Place the model of the jack you constructed for Activity 4 of Investigation 1 (page 387) on a coordinate system. Fix point *A* at the origin and position side *AE* along the positive *x*-axis. Collect data on the coordinates of points *C* and *B* as point *C* moves away from point *A* along the *x*-axis.

   a. As point *C* moves, what is the relationship between the *x*-coordinate of point *C* and the *x*-coordinate of point *B*?

   b. What is the relationship between the *y*-coordinate of point *D* and the *y*-coordinate of point *B*, for each position of point *C*? Explain your reasoning.

   c. Make a scatterplot of the (*x-coordinate of C*, *y-coordinate of B*) data. What pattern do you see? Does a linear, exponential, or power model seem to fit these data?

   d. Make a scatterplot of the (*x-coordinate of B*, *y-coordinate of B*) data. What pattern do you see? Does a linear, exponential, or power model seem to fit these data?

4. Ramps often provide alternate routes for people who are not able to use stairs. The design of a ramp is not based solely on the smoothness of its surface.

a. The maximum *gradient* of ramps for handicap access is 8.33%. (That is, the slope is 0.0833 or $\frac{5}{60}$.) The maximum ground distance allowed for a single ramp at the maximum gradient is 30 feet. Design two different accesses to a door which has its sill 3.5 feet from the ground surface.

b. The *mechanical advantage* of a single ramp is the ratio $\frac{\text{effort distance}}{\text{resistance distance}}$, that is, $\frac{\text{height of ramp}}{\text{length of ramp}}$. Which single ramp in your response to Part a has the better (lower) mechanical advantage?

c. What is the minimum length ramp that can be used in this situation? What is the mechanical advantage of that ramp?

Reflecting

1. People who travel often need an easy way to move their luggage. The cart below will fold up to a small size for storage.

BEST CARRY-ON LUGGAGE CART

This cart was ranked best against other models tested by a panel of our customers for its superior stability, design features and overall performance. Exceptionally versatile, its four-wheeled design means that you can push or pull up to 225 pounds of luggage with remarkable stability and without putting any weight-strain on your arms.

Hammacher Schlemmer

a. Explain how a triangle with a variable-length side is used in this cart's design.

b. Design your own carry-on luggage cart that does not make use of a triangle with a variable-length side. Illustrate your design with a sketch and compare its functionality to the cart in the photo.

2. Choose one of the following items and explain how it makes use of a triangle with a side of adjustable length.

a. Umbrella

b. Music stand

c. Window-opening apparatus

d. Slide projector (adjustable)

e. Backhoe

f. Metal folding chair

g. Pruning shears

3. Parallelogram linkages and variable-sided triangle linkages each have a number of uses. Which of the two linkages do you consider to be the most useful? Why?

4. Listening to engineers, you may hear the statement, "form follows function." What do you think they mean by this statement? To what extent do you agree with it?

Extending

1. A large hoisting derrick has a boom *AB* that is 15 meters long. It is raised by a cable which may be shortened using a winch at three rates: $\frac{1}{4}$ meter per second, $\frac{1}{3}$ meter per second, or $\frac{1}{2}$ meter per second.

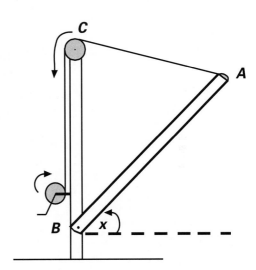

   a. What is the variable side in this situation?

   b. Assume the boom *AB* begins in a horizontal position and *BC* is 15 meters. For each rate, make a graph showing the relationship between time and the measure of the *angle of elevation x*, until the boom is brought to a vertical position. Describe the pattern of change.

   c. Again assume that boom *AB* begins in a horizontal position and *BC* is 15 m. For each rate, make a graph of data showing the relationship between time and the distance from point *A* to the horizontal dashed line, until the boom is brought to a vertical position. Describe the pattern of change in the graph.

2. Suppose a cold frame (see Modeling Task 1) has been designed to be held open by a prop 30 cm long, attached to the top 50 cm from the hinged end. The free end of the prop rests in one of several notches, spaced equally apart, on the side of the cold frame. Imagine or sketch the prop placed in any notch on the side.

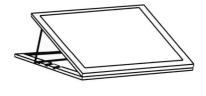

   a. If the free end of the prop is moved to a notch *farther* from the hinged end of the frame, what happens to the measure of the angle at the hinged end? Is the size of the change constant as the prop is moved *out* to each notch?

b. If the free end is returned to its starting position and then moved to a notch *closer* to the hinged end of the box, what happens to the measure of the angle at the hinge? Is the size of the change constant as the prop is moved *in* to each notch?

c. To make the largest opening for this cold frame, where should the designer position the notch? With this largest opening, what is the angle formed by the prop where it meets the side of the cold frame?

d. For what kind of hinged structure might you want to design a prop that attaches near the hinged end, rather than in the middle of the structure? Which attachment place do you think works best for the cold frame?

e. Now experiment with changing either the length of the prop or where it is attached to the raised lid. Under what conditions will there be two possible ways to place the prop to get a desired opening size? Under what conditions will there be only one way to get a desired opening size?

3. A ramp is an example of an *inclined plane*. By adjusting the height of an incline you can investigate basic scientific principles.

a. Imagine placing a skateboard on a ramp and allowing it to roll with no initial push. What do you think is the relationship between the height of the ramp and the elapsed time for the skateboard to roll down it?

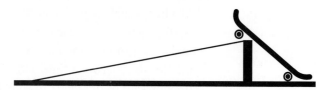

b. Design and conduct an experiment to investigate the relationship between height of an inclined plane and time for an object to roll down the plane. (The science department in your school may have materials to help in the design of your experiment.)

c. Find a rule that you believe fits the pattern in the (*height, elapsed time*) data that you have collected.

d. Write a report summarizing your experiment's design, methods, and results. Describe any limitations of your modeling rule.

4. Refer to the inclined-plane experiment in Task 3 above. Design and conduct an experiment to investigate how the weight of the object rolled down the ramp affects the pattern of change in the corresponding (*height, elapsed time*) data. How is the effect of weight seen in the scatterplots of the data? In the corresponding modeling equations?

INVESTIGATION 2 What's the Angle?

In the previous investigation, you learned that when the rigidity property of triangles is combined with the ability to adjust the length of a side, the opportunities for useful application expand greatly. You probably noticed that the methods used to determine lengths and angle measures involved measuring the models you made. In this investigation, you will use right triangles and similarity to explore other ways in which lengths and angle measures can be determined.

Recall that if two figures are *similar* with a scale factor of *s*, then corresponding angles are congruent (≅) and lengths of the corresponding sides are related by the multiplier *s*. The two school crossing signs shown below are similar.

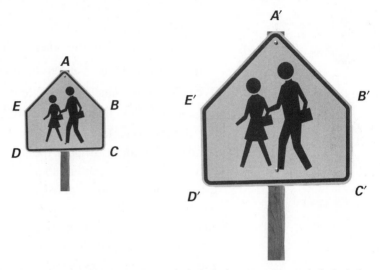

Here pentagon *ABCDE is similar to A′B′C′D′E′* (*ABCDE ~ A′B′C′D′E′*) with a scale factor of 2. ∠*B* corresponds to ∠*B′*, so ∠*B* ≅ ∠*B′*. (∠*B is congruent to* ∠*B′*.) Segment *ED* corresponds to segment *E′D′*, so 2 · *ED* = *E′D′* or, equivalently, $ED = \frac{E'D'}{2}$.

1. Imagine that you and a classmate each draw a triangle with three angles of one triangle congruent to three angles of the other triangle. Do you think the two triangles will be similar? Make a conjecture.

   a. Now conduct the following experiment. Have each member of your group draw a segment (no two with the same length). Use a protractor to draw a 50° angle at one end of the segment. Then draw a 60° angle at the other end of the segment to form a triangle.

   ■ What should be the measure of the third angle? Check your answer.

   ■ Are these triangles similar to one another? What evidence can you give to support your view?

   b. Repeat Part a with angles measuring 40° and 120°. Are these triangles similar? Give evidence to support your claim.

c. Which, if any, of the following statements do you think are always true? Justify your response with reasons or a counterexample.

- If one triangle has three angles congruent to three corresponding angles on another triangle, then the triangles are similar.

  - If one triangle has two angles congruent to two corresponding angles on another triangle, then the triangles are similar.

  - If one triangle has one angle congruent to one angle on another triangle, then the triangles are similar.

2. Now apply your discoveries in Activity 1 to the special case of right triangles.

a. Each group member should draw a segment *AC* (each a different length). Using your segment *AC* as a side, draw △*ABC* with ∠*A* measuring 35° and a right angle at *C*. It is important to draw your triangle very carefully. What is the measure of the other angle (∠*B*)?

b. Are the triangles your group members drew similar? Explain.

c. Choose the smallest triangle drawn in Part a. Determine the approximate scale factors relating this triangle to the others drawn by group members.

For a right triangle *ABC*, it is standard procedure to label the right angle with the capital letter *C* and to label the sides opposite the three angles lower case *a*, *b*, and *c* as shown. Complete a labeling of your triangle in this way. (Additional ways to refer to the sides of a right triangle are

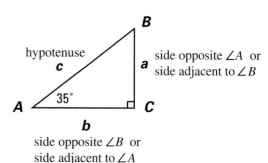

also included in the diagram. The **hypotenuse** is always the side opposite the right angle, but the designation of the other sides depends on which angle is considered.)

3. A diagram of a right △*ABC* is given below. Give the measures of the following angles or sides:

a. ∠*C*

b. ∠*B*

c. Side opposite ∠*A*

d. Side (leg) adjacent to ∠*B*

e. Side (leg) adjacent to ∠*A*

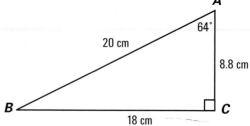

4. Refer to the right triangles your group drew for Activity 2.

  a. Make a table like the one below. Each group member should choose a unit of measure. Then carefully measure and calculate the indicated ratios for the right $\triangle ABC$ that you drew. Express the ratios to the nearest 0.01. Investigate patterns in the three ratios for your group's triangles.

Right Triangle Side Ratios

Ratio	Student 1	Student 2	Student 3	Student 4
$\dfrac{a}{c}$				
$\dfrac{b}{c}$				
$\dfrac{a}{b}$				

  b. Compare the ratios from your group with those of other groups.

  c. On the basis of Parts a and b, make a conjecture about the three ratios in the table for any right $\triangle ABC$ with a 35° angle at A.

  d. How could the three ratios be described in terms of the hypotenuse and the sides opposite and adjacent to $\angle A$? In terms of the hypotenuse and the sides opposite and adjacent to $\angle B$?

  e. Make a conjecture about the three ratios in the table for any right triangle with a 35° angle.

  f. Make a conjecture about the three ratios for any right triangle with a 55° angle.

5. As a group, draw several examples of a right $\triangle ABC$ in which $\angle A$ has a measure of 40° and $\angle C$ has a measure of 90°.

  a. Compute the three ratios $\dfrac{a}{c}$, $\dfrac{b}{c}$, and $\dfrac{a}{b}$. Record the ratios in a table like the one in Activity 4.

  b. What pattern do you see in these ratios?

  c. How is the pattern for these ratios similar to the pattern for the ratios in Activity 4? How is it different?

  d. What seems to cause the differences in the results from the two activities? Test your conjecture by experimenting with another set of similar right triangles and inspecting the ratios.

You have observed that as the measure of an acute angle changes in a right triangle, ratios of the lengths of the sides also change. In fact, each ratio is a function of the size of the angle. These relationships are important because they relate measures of angles (in degrees) to ratios of linear measures (in centimeters, miles, and so on). The relationships or functions have special names. For a right triangle ABC with sides a, b, and c, the **sine**, **cosine**, and **tangent** ratios for $\angle A$ are defined as follows:

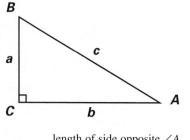

$$\text{sine of } \angle A = \sin A = \frac{\text{length of side opposite } \angle A}{\text{length of hypotenuse}} = \frac{a}{c}$$

$$\text{cosine of } \angle A = \cos A = \frac{\text{length of side adjacent to } \angle A}{\text{length of hypotenuse}} = \frac{b}{c}$$

$$\text{tangent of } \angle A = \tan A = \frac{\text{length of side opposite } \angle A}{\text{length of side adjacent to } \angle A} = \frac{a}{b}$$

These ratios are called **trigonometric ratios**.

Sine B, cosine B, and tangent B are similarly defined by forming the ratios using the sides opposite and adjacent to $\angle B$. The abbreviations are $\sin B$, $\cos B$, and $\tan B$.

6. Refer to the triangles you drew for Activity 5. Write the definitions for $\sin B$, $\cos B$, and $\tan B$, and then find $\sin 50°$, $\cos 50°$, and $\tan 50°$.

7. Suppose you have a right triangle with an acute angle of $27.5°$. One way you could find the sine, cosine, and tangent of $27.5°$ would be to make a very accurate right triangle and measure. Another way is to use your calculator.

   a. To calculate a trigonometric ratio for an angle measured in degrees, first be sure your calculator is set in *degree* mode. Then simply press the keys corresponding to the ratio desired. For example, to calculate $\sin 27.5°$ on most graphing calculators, press ⌜SIN⌝ 27.5 ⌜ENTER⌝. Try it. Then calculate $\cos 27.5°$ and $\tan 27.5°$.

   b. Compare $\sin 27.4°$ and $\sin 27.6°$ to your value for $\sin 27.5°$. How many decimal places should you include to show that the angle whose sine you are finding was measured to the nearest $0.1°$?

   c. How many decimal places should you report for $\sin 66.5°$ to indicate that the angle was measured to the nearest $0.1°$?

   d. Use your calculator to find the sine, cosine, and tangent of $35°$ and of $50°$. Compare these results with those you obtained by measuring in Activities 4 through 6.

You can use your calculator to compute values for sine, cosine, and tangent of any angle. Several hundred years ago mathematicians spent years calculating these ratios by hand to several decimal places so that they could be used in surveying and astronomy. Until recently, before scientific and graphing calculators became common, people usually looked up the ratio values from a large table. Now that a calculator replaces this tedious work, you can concentrate on understanding trigonometric ratios and their uses.

Checkpoint

Knowing some information about a triangle or pair of triangles often allows you to conclude other information.

a What can you conclude about two triangles if two angles of one are congruent to two angles of the other?

b Why does knowing the measure of an acute angle of a right triangle completely determine the triangle's shape?

c For two different triangles ABC and DEF, in which $\angle A$ and $\angle D$ both have measure $x°$ and $\angle C$ and $\angle F$ are right angles, what can you say about the ratios $\frac{AC}{AB}$ and $\frac{DF}{DE}$? Explain why this makes sense.

d If $\sin x° = \frac{b}{c}$ in right $\triangle ABC$, then which angle has measure $x°$? Which angle has measure $x°$ when $\cos x° = \frac{b}{c}$?

Be prepared to discuss your responses with the entire class.

▶ On Your Own

Refer to the drawing of $\triangle PQR$ below.

a. Which is the side opposite $\angle P$? The side adjacent to $\angle R$? The hypotenuse?

b. Use the Pythagorean Theorem to find the length of side PQ.

c. Find the following trigonometric ratios:

- $\sin P$
- $\sin R$
- $\tan P$
- $\cos R$

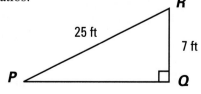

d. Valerie measured $\angle R$ and found it to be almost 74°. Use a calculator and your results from Part c to estimate the measure of $\angle R$ to the nearest 0.1°.

INVESTIGATION 3 Measuring Without Measuring

Shown below is Chicago's *Bat Column*, a sculpture by Claes Oldenburg.

1. In your group, brainstorm about possible ways to determine the height of the sculpture.

   a. Choose one method and write a detailed plan.

   b. Trade plans with another group and compare the two plans.

   c. What assumptions did the other group make in devising its plan?

   d. Which plan seems easier to carry out? Why?

Your class probably thought of several plans to determine the height of *Bat Column*. For example, one could use an extension ladder on a fire truck to climb to the top and drop a weighted and measured cord to the ground. This would be a *direct measurement* procedure.

2. An *indirect* way to measure the height of *Bat Column* would be to use a right triangle and a trigonometric ratio.

   a. In the situation depicted below, what lengths and angles could you determine easily by direct measurement (and without using high-powered equipment)?

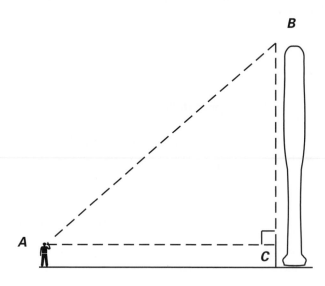

   b. Which trigonometric ratios of ∠*A* involve side *BC*? Of these, which also involve a measurable length?

c. Which of the trigonometric ratios of ∠B involve side *BC* and a measurable length? If you know the size of ∠A, how can you find the measure of ∠B?

d. Krista and D'wan decided to find the height of *Bat Column* themselves. First Krista chose a spot to be point *A*, 20 meters from the sculpture (point *C*). D'wan used a *clinometer*, like the one shown at the right, to estimate the measure of ∠A (the *angle of elevation* from the horizontal to the top of the bat). He measured ∠A to be 55°. What is the measure of ∠B?

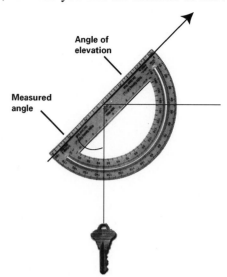

e. Krista and D'wan proceeded to find the height of the bat independently as shown below.

D'wan

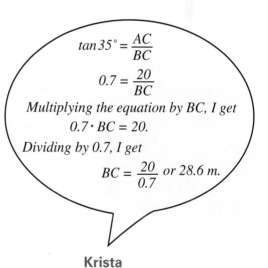

I need to find BC so that
$$\frac{BC}{AC} = \tan 55°.$$
But tan 55° = 1.43 and AC = 20 m. So I need to solve
$$\frac{BC}{20} = 1.43.$$
If I multiply the equation by 20, I get
$$BC = 1.43 \cdot 20$$
$$BC = 28.6 \ m$$

$$\tan 35° = \frac{AC}{BC}$$
$$0.7 = \frac{20}{BC}$$
Multiplying the equation by BC, I get
$$0.7 \cdot BC = 20.$$
Dividing by 0.7, I get
$$BC = \frac{20}{0.7} \ or \ 28.6 \ m.$$

Krista

- Analyze D'wan's thinking. Why did he multiply by 20?
- Analyze Krista's thinking. Why did she multiply by *BC*? Why did she divide by 0.7?
- Are the answers correct? Explain your response.

f. How could you use Krista's and D'wan's work to help estimate the height of *Bat Column*?

g. Kim said he could find the length *AB* (the line of sight distance) by solving $\cos 55° = \frac{AC}{AB}$. Analyze Kim's thinking shown here.

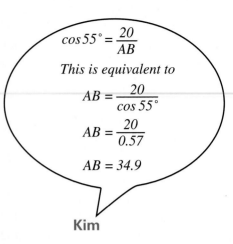

$$\cos 55° = \frac{20}{AB}$$

This is equivalent to

$$AB = \frac{20}{\cos 55°}$$

$$AB = \frac{20}{0.57}$$

$$AB = 34.9$$

Kim

- Explain Kim's thinking.

- Is Kim correct?

- What is another way Kim could have found *AB* using trigonometric ratios?

- Could you find *AB* without using trigonometric ratios? Explain your reasoning.

3. Each part below gives data for right △*ABC*. Sketch a model triangle and then, using your calculator, find the lengths of the remaining two sides.

a. ∠*B* = 52°, *a* = 5 m   **b.** ∠*A* = 78°, *a* = 5 mi

c. ∠*A* = 21°, *b* = 8 in.   **d.** ∠*B* = 8°, *b* = 8 ft

e. ∠*B* = 37°, *c* = 42 yd   **f.** ∠*A* = 82°, *c* = 14 cm

4. Terri is flying a kite and has let out 500 feet of string. Her end of the string is 3 feet off the ground.

a. If ∠*KIT* has a measure of 40°, approximately how high off the ground is the kite?

b. As the wind picks up, Terri is able to fly the kite at a 56° angle with the horizontal. Approximately how high is the kite?

c. What is the highest Terri could fly the kite on 500 feet of string? What would be the measure of ∠*KIT* then?

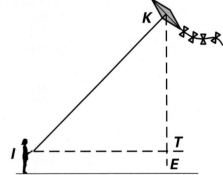

d. Experiment with your calculator to estimate the measure of ∠*KIT* needed to fly the kite at a height of 425 feet.

In the previous situations, you used trigonometric ratios to determine an unknown or inaccessible distance. In Activity 4 Part d, you probably found a way to find the measure of an angle when you know the lengths of two sides in a right triangle.

5. Estimate (to the nearest degree) the measure of acute angle B for each of the following trigonometric ratios of $\angle B$. Check your estimate in each case by drawing a model right $\triangle ABC$, using sides whose lengths give the appropriate ratio, and then measuring $\angle B$.

a. $\sin B = \frac{3}{5}$

b. $\cos B = \frac{1}{2}$

c. $\tan B = \frac{4}{5}$

d. Consider how you found the measure of an acute angle when you knew a trigonometric ratio for a right triangle with that angle measure. Compare your group's approach with the approaches of other groups.

6. You know how to use a calculator to produce a trigonometric ratio when you know the measure of an angle. You also can use a calculator to produce the angle when you know a trigonometric ratio as in Activity 5.

a. Suppose you know $\sin A = \frac{4}{5} = 0.8$. Use the "$\sin^{-1}$" function of your calculator to compute the angle whose sine is 0.8. (Make certain your calculator is set in degree mode.)

```
sin⁻¹(.8)
            53.13010235
```

b. What would you get if you calculated the sine of the angle in the calculator display at the right?

c. Use your calculator to find the measure of $\angle B$ that corresponds to each of the ratios given in Activity 5. Compare these values to the values you obtained in that activity.

d. Use your calculator to find the measure of the angle in each of the following cases.

■ $\tan B = 1.84$

■ $\sin A = 0.852$

■ $\cos B = 0.213$

7. The Canadian National Tower in Toronto, Ontario, is approximately 553 meters tall. This tower is the tallest free-standing structure in the world.

a. Sketch the tower and add the features described in Parts b, c, and d as you work to answer each part.

b. At some time on a sunny day, the sun makes the tower cast a 258-meter shadow. What is the measure of the angle formed by a sun ray and the ground at the tip of the shadow?

c. From the top of the Canadian National Tower, a boat is observed in Lake Ontario, approximately 8,000 meters away from the base of the tower. Assume the base of the tower is approximately level with the lake surface. What angle below the horizontal must the observer look to see the boat?

d. Estimate the line of sight distance from the observer to the boat in Part c. Find this distance using trigonometric ratios and without using them.

8. Lakeshia is about 1.7 meters tall. When standing 5 meters from her school building, her angle of sight to the top of the building is 75°.

a. Estimate the height of the building.

b. Suppose Lakeshia moves to a position 10 meters from the building. What is the angle of her new line of sight to the top of the building?

c. Marcio, who is also about 1.7 meters tall, is standing on top of the building. He sees Lakeshia standing 15 meters from the building. At what angle below the horizontal is his line of sight to Lakeshia?

Checkpoint

Trigonometric ratios are useful to calculate lengths and angles in right-triangle models. Refer to the right triangle shown below in summarizing your thinking about how to use trigonometric ratios.

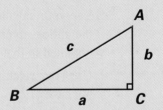

ⓐ If you knew a and the measure of $\angle B$, how would you find b? What calculator keystroke sequence would you use?

ⓑ If you knew b and c, how would you find the measure of $\angle A$? What calculator keystroke sequence would you use?

ⓒ If you knew b and the measure of $\angle B$, how would you find c? What calculator keystroke sequence would you use?

Be prepared to explain your methods to the whole class.

The **angle of elevation** to the top of an object is the angle formed by the horizontal and the line of sight to the top of the object. In the diagram below, $\angle ACD$ is the angle of elevation. The **angle of depression** to an object is the angle formed by the horizontal and the line of sight to the object below. In the diagram, $\angle BAC$ is the angle of depression.

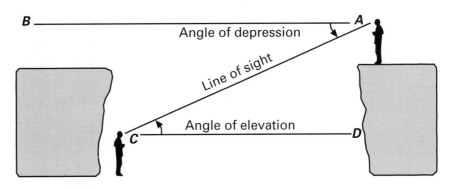

On Your Own

A person on an oil-drilling ship in the Gulf of Mexico sees a semi-submersible platform with a tower on top of it. The tower stands 130 meters above the platform floor.

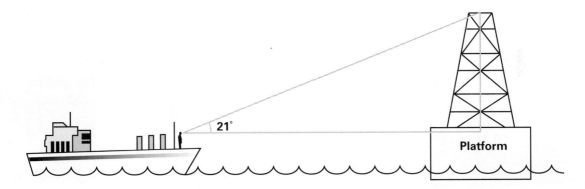

a. If the observer's position on the boat is 15 meters under the floor of the platform and the angle of elevation to the top of the rig is 21°, what three distances can you find? Find them.

b. Suppose the boat moves so that it is 200 meters from the center of the oil rig. What is the angle of elevation now?

Modeling

1. At places along the south bank of the Kalamazoo River, the river is substantially below the level of the ground. The land slopes downward at an angle of approximately 63° from the horizontal. The best way to get to the water is to go directly down the slope to the water's edge, a distance of approximately 123 meters.

    a. Make a sketch of this situation showing distances and angles.

    b. How far, horizontally, is the edge of the water from the edge of the bank, just where the land begins to slope?

    c. How far above the surface of the Kalamazoo River is the land on the south bank?

2. Commercial aircraft usually fly at an altitude between 9 and 11 kilometers (29,000 to 36,000 feet). When landing, their gradual descent to an airport runway occurs over a long distance.

    a. Suppose a commercial airliner begins its descent from an altitude of 9 km with an angle of descent of 2.5°. At what distance from the runway should the descent begin?

    b. Suppose a commercial airliner flying at an altitude of 11 km begins its descent at a horizontal distance 270 km from the runway. What is its angle of descent?

    c. The *cockpit cutoff angle* of an airliner is the angle from the horizontal to the line of sight between the pilot and the nose of the plane. Suppose a pilot is flying an aircraft with a cockpit cutoff angle of 14° at an altitude of 1.5 km. How far from the near edge of a lake, measuring along her line of sight, will she be when she is last able to see that edge of the lake?

    d. What is the horizontal distance to the lake when the pilot in Part c is last able to see the edge of the lake?

Cockpit cutoff angle

3. Astronomers use the length of shadows in craters on the moon's surface to determine the depth of the craters. Since they can determine the positions of the sun and moon rather precisely, they can compute the angle of elevation of the sun from the crater floor.

a. Sketch a side view of a crater, floor, and shadow.

b. What trigonometric ratio can be used to estimate the height of the crater wall?

c. The shadow in one crater is 315 meters when the angle of elevation to the sun is 24°. Estimate the height of the crater wall.

d. To check the result of Part c, astronomers identified a shadow of 264 meters when the angle of elevation was 28°. Is this consistent with the data in Part c?

4. The *grade* of a road is the ratio of the vertical distance traveled to the horizontal distance traveled. The angle of incline is the angle between the horizontal and the roadbed.

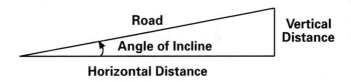

a. To find the grade of a roadbed, a level sighting instrument is used to help adjust the heights of two vertically-placed extendible rods so that their upper tips are on the horizontal line of sight. Make a sketch of the terrain shown below, and then sketch in the rods *AC* and *BD* so that $\overline{CD}$ is horizontal.

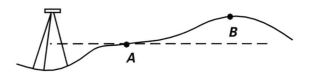

b. What measures in the diagram of Part a would you need to know to determine the grade of the roadbed?

c. Suppose the rod at *A* is 2.5 meters, the rod at *B* is 1.2 meters, and the distance from *C* to *D* is 54 meters. Find the grade.

d. Find the angle of incline for the situation in Part c. What trigonometric ratio did you use?

e. Suppose the road grade is 7%. What is the angle of incline?

f. An exit ramp for interstate highway I-94 is to be built with a 4.34% grade to an overpass that is 11 meters above the road surface. How far horizontally from the overpass should the engineers plan to begin the ramp?

g. How much could the ramp be shortened if a 5% grade were used?

Organizing

1. Two right triangles, *ABC* and *A'B'C'*, with right angles at *C* and *C'*, are similar with a scale factor of *k*. The lengths of the sides of △*ABC* are *a*, *b*, and *c*.

   a. Write expressions for the lengths of the sides of △*A'B'C'*.

   b. Write expressions for sin *A*, sin *A'*, tan *A*, and tan *A'*. What do you notice?

   c. On the basis of your observations in Parts a and b, write an argument demonstrating that cos *A* = cos *A'* for any two similar right triangles *ABC* and *A'B'C'*.

2. Enter the numbers 0 through 90, in steps of 5, into List 1 of your calculator or computer software. If you are using a graphing calculator, make certain it is set in degree mode. If you are using computer software, find out how to make certain the software uses angle measurements expressed in degrees.

   a. Fill List 2 with the sine of the angle measurements in List 1.

   ■ As the measure of an angle increases from 0° to 90°, how does the sine ratio change?

   ■ Examine a scatterplot of this data. How does this plot compare with those of linear, exponential, power, and quadratic models?

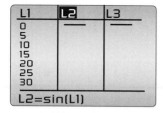

   b. Repeat Part a, filling List 3 with the cosine of the angle measurements in List 1.

   c. Delete the 90 from List 1 and repeat Part a, filling List 4 with the tangent of the angle measurements in List 1.

3. Systematically vary the acute angles of a right $\triangle ABC$ with $\angle C$ a right angle, and investigate possible patterns relating the trigonometric ratios given below in Parts a–d. It may help to modify the method outlined in Organizing Task 2, with numbers 5 through 85, in steps of 5, in List 1. The entries in List 1 would be possible degree measures for $\angle A$, and you can place in List 2 the corresponding degree measures of $\angle B$.

   a. $\sin A$ and $\cos B$

   b. $\tan A$ and $\tan B$

   c. $\dfrac{\sin A}{\cos A}$ and $\tan A$

   d. $(\sin A)^2$ and $(\cos A)^2$

4. Use the definitions of the trigonometric ratios to explain why the patterns you observed in Parts c and d of Organizing Task 3 will be true for any acute angle A in a right triangle.

5. The Pyramid of Cheops in Egypt is a right square pyramid. The base edge measures about 230 meters and each face makes an angle with the horizontal desert floor of 51.8°.

   a. Make a model or a sketch of the Pyramid of Cheops. (Use coffee-stirrers or other manipulatives for your model.)

   b. Determine the height of the pyramid.

   c. If you were climbing the pyramid, what would be the shortest route to the top? What is the length of this route?

   d. What is the *grade* of a pyramid face? (See Modeling Task 4 on page 407 for a definition of *grade*.)

   e. Determine the dimensions of the faces of the pyramid.

   f. Determine the volume of the Pyramid of Cheops using the formula $V = \frac{1}{3} Bh$, where B is the area of the base and h is the height of the pyramid.

Reflecting

1. In this lesson, as well as in previous units, you have engaged in important kinds of mathematical thinking. From time to time, it is helpful to examine the kinds of thinking that are broadly useful in doing mathematics. Look over the three investigations and the MORE tasks you completed in this lesson, and consider some of the mathematical thinking you have done. Describe examples in which you did each of the following:

   a. Experimented

   b. Used a variety of representations, like tables, graphs, equations, or verbal descriptions

   c. Searched for patterns

   d. Formulated or found a mathematical model

   e. Visualized

   f. Made and checked conjectures

2. For a right $\triangle ABC$ with $\angle C$ a right angle, if the sides are measured in feet, what is the unit of measure for $\cos A$? Explain your reasoning. What if one side is measured in inches and the other two sides are measured in feet?

3. Consult resources on the history of mathematics to determine the origin of the words sine, cosine, and tangent.

4. How is the slope of a line $y = ax$, where $a > 0$, related to the tangent of the angle formed by the line and the positive x-axis?

   a. Write an equation of the line through the origin that forms a 60° angle with the positive x-axis.

   b. Determine the measure of the angle formed by the line $y = 4x$ and the positive x-axis.

5. Find the tallest object on your school campus that cannot easily be measured directly. Prepare a plan to determine the height of the object. Carry out your plan, if it is reasonably possible.

Extending

1. In $\triangle ABC$ shown here, $\overline{BD}$ is an altitude.

   a. Write expressions to represent $\sin A$ and $\sin C$.

   b. Use the results of Part a to write a single equation involving $\sin A$, $\sin C$, a, and c.

   c. Repeat Parts a and b for the altitude from C to $\overline{AB}$, using $\angle A$ and $\angle B$.

   d. What can you conclude about $\frac{\sin A}{a}$, $\frac{\sin B}{b}$, and $\frac{\sin C}{c}$?

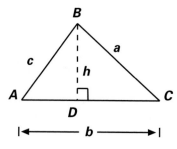

2. Refer to the diagram for Extending Task 1. Explain why the following is true: *Area of* $\triangle ABC = \frac{1}{2} ab \cdot \sin C$

3. While camping by the Merced River in Yosemite Valley, a group of friends was admiring a particular tree on the opposite bank. Maria claimed that the height of the tree could be determined from the group's side of the river by the following method:

- Measure a 50-meter segment, $\overline{AB}$, on this shore.
- Considering the tree to be located at point *C*, measure $\angle BAC$ and $\angle ABC$.
- From *A*, measure the angle of elevation to the top of the tree.
- Use the measurements above with some trigonometric ratios to calculate the height.

The friends measured $\angle BAC$ to be 54° and $\angle ABC$ to be 74°. The angle of elevation from point *A* to the treetop was 11°.

Draw a sketch of this situation and determine whether Maria was correct. If she was, compute the height; if she was not, explain why not.

4. A *clinometer* (see page 401) is an instrument used to measure angles of elevation.

    a. Consult references in the library to learn how a clinometer works.

    b. Build a wooden or stiff cardboard model of a clinometer.

    c. Use your model to estimate the height of a landmark in your community. Check your estimate against whatever information might be recorded in the community.

5. It is unfortunate, but custodians for public buildings often need to clean graffiti off walls. Mr. Pyper, the custodian at Downhill High School, has a 20-foot ladder. According to information on the ladder, it is safe at angles of 75° or less. If Mr. Pyper can stand on the rung 2 feet below the top of the ladder and reach $7\frac{1}{2}$ feet with one arm, what is the maximum height he can reach using the ladder?

6. Use the clinometer you designed in Extending Task 4. Choose a landmark in your community. Move to a location 100 paces from the landmark, and then measure and record the angle of elevation. Move two paces closer to the landmark and repeat the procedure. Complete a table of data pairs (*paces from landmark, angle of elevation*) for 30 measurements. What patterns do you see in the data? Describe the patterns you see in the rate of change in the angles. Make a scatterplot of the (*paces from landmark, angle of elevation*) pairs and describe the graph.

The Power of the Circle

Quadrilaterals and triangles are used to make everyday things work. Right triangles are the basis for trigonometric ratios relating angle measures to ratios of lengths of sides. Another family of shapes that is broadly useful is the circle and its three-dimensional relatives, such as spheres, cylinders, and circular cones.

An important characteristic of a circle is that it has rotational symmetry about its center. For example, the hub of an automobile wheel is at the center of a circle. As the car moves, it travels smoothly because the circular tire keeps the hub a constant distance from the pavement.

Motors often rotate a cylindrical drive shaft. The more energy output, the faster the drive shaft turns. On an automobile engine, for example, a belt connects three pulleys, one on the crankshaft, one which drives the fan, and another which drives the alternator. When the engine is running, the fan cools the radiator while the alternator generates electrical current.

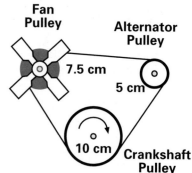

Fan Pulley
Alternator Pulley
7.5 cm
5 cm
10 cm
Crankshaft Pulley

Think About This Situation

The diameter measurements given in the diagram above are for a particular four-cylinder sports car.

a How does the speed of the crankshaft affect the speed of the fan? Of the alternator?

b The idle speed of the crankshaft of a four-cylinder sports car is about 850 rpm (revolutions per minute).

■ How far, in centimeters, would a point on the edge of the fan pulley travel in one minute?

■ Do you think a similar point on the connected alternator pulley would travel the same distance in one minute? Why or why not?

c Describe another situation in which turning one pulley (or other circular object) turns another.

INVESTIGATION 1 Follow That Driver!

In this investigation, you will explore how rotating circular objects that are connected can serve useful purposes. A simple way to investigate how the turning of one circle (the *driver*) is related to the turning of another (the *follower*) is to experiment with thread spools and rubber bands. The spools model the pulleys, sprockets, or gears; the rubber bands model the belts or chains connecting the circular objects. Complete the first two activities working in pairs.

1. Use two thread spools of different sizes. If you use spools that still have thread be certain the thread is securely fastened.

   Put each spool on a shaft, such as a pencil, which permits the spool to turn freely. Make a *driver/follower* mechanism by slipping a rubber band over the two spools and spreading the spools apart so the rubber band is taut enough to reduce slippage. Choose one spool as the driver.

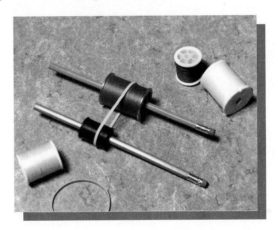

   a. Turn the driver spool. Describe what happens to the follower spool.

   b. Turn the driver spool one complete revolution. Does the follower spool make one complete revolution, or does it make more or fewer turns?

   c. Turn the driver spool one complete revolution. How far does the rubber band advance?

   d. Design and carry out an experiment that gives you information about how turning the driver spool affects the amount of turn of the follower spool, when the spools have different *radii*. Use whole number and fractional turns of the driver. Organize your (*driver turn amount, follower turn amount*) data in a table.

   e. Plot your (*driver turn amount, follower turn amount*) data. Find an algebraic model that fits the data.

   f. Compare your scatterplot and model with those of other pairs of students.

   ■ How are they the same? How are they different?

   ■ What might explain the differences?

g. Examine the driver/follower spool data below.

- What pattern would you expect to see in a plot of these data?
- What algebraic model do you suspect would fit these data?

Driver/Follower Data Set 1	
Driver Radius: 2.5 cm	**Follower Radius: 2 cm**
Driver Turn Amount (in revolutions)	**Follower Turn Amount (in revolutions)**
0.5	0.6
1	1.3
2	2.5
3	3.8
5	6.2
8	10.0
10	12.5
12	15.0

Driver/Follower Data Set 2	
Driver Radius: 1 cm	**Follower Radius: 1.5 cm**
Driver Turn Amount (in revolutions)	**Follower Turn Amount (in revolutions)**
0	0.0
1	0.7
3	2.0
5	3.4
7	4.6
9	6.0
12	8.0
15	10.0

2. Reverse the driver/follower roles of the two spools. How does turning the driver affect the follower now?

3. Suppose the driver spool has a radius of 2 cm and the follower spool has a 1-cm radius.

a. If the driver spool turns through 90 degrees, through how many degrees will the follower spool turn? Support your position experimentally or logically.

b. In general, how will turning the driver spool affect the follower spool? Provide evidence that your conjecture is true, using data or reasoning about the situation.

c. How do the lengths of the radii of the spools affect the driver/follower relation? Answer as precisely as possible.

d. The number by which the turn or speed of the driver is multiplied to get the turn or speed of the follower is often called the **transmission factor** from driver to follower. What is the transmission factor for a driver with a 4-cm radius and a follower with a 2-cm radius? If you reverse the roles, what is the transmission factor?

e. List two sets of driver/follower spool radii so that each set will have a transmission factor of 3. Do the same for transmission factors of $\frac{3}{2}$ and $\frac{4}{5}$.

f. If the driver has radius r_1 and the follower has radius r_2, what is the transmission factor?

4. In addition to designing a transmission factor into a pulley system, you must also consider the directions in which the pulleys turn.

    a. In the spool/rubber band systems you made, did the driver and follower turn in the same direction? Sketch a spool/rubber band system in which turning the driver clockwise turns the follower clockwise also. Label the driver and follower. How would this system look if both the driver and follower were to turn counterclockwise?

    b. Sketch a spool/rubber band system in which turning the driver spool clockwise turns the follower counterclockwise. Make a physical model to check your thinking.

    c. Suggest a way to describe the transmission factor for the system in Part b. Should the transmission factor differ from a system using the same spools turning in the same direction? If so, how? If not, why not?

5. A clockwise driver/follower system has a driver with a 5-inch radius and a follower with a 4-inch radius.

    a. Sketch the system. What is the transmission factor for this system?

    b. What are the circumferences of the two pulleys? How could you use the lengths of the circumferences to determine the transmission factor of the system? Explain why this is reasonable.

    c. How far does a point on the edge of the driver travel in one revolution of the driver?

    d. If the driver is rotating 50 revolutions per minute (rpm), how far does the point in Part c travel in 1 minute?

    e. If the driver is turning at 50 rpm, how fast is the follower turning?

    f. In one minute, how far does a point on the circumference of the follower travel? Compare this result with that in Part d and explain your findings.

6. Wanda is riding her mountain bicycle using the crankset (also called the pedal sprocket) with 42 teeth of equal size. The rear-wheel sprocket being used has 14 teeth of a size equal to the crankset teeth.

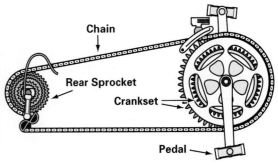

a. What does the "teeth per sprocket" information tell you about the circumferences of the two sprockets? Translate the information about teeth per sprocket into a transmission factor.

b. Suppose Wanda is pedaling at 80 revolutions per minute.

- What is the rate at which the rear sprocket is turning? Explain.

- What is the rate at which the rear wheel is turning? Explain.

c. The wheel on Wanda's mountain bike has a radius of about 33 cm.

- How far does the bicycle travel for each complete revolution of the 14-tooth rear sprocket?

- How far does the bicycle travel for each complete revolution of the front sprocket?

- If Wanda pedals 80 rpm, how far will she travel in one minute?

d. How long will Wanda need to pedal at 80 rpm to travel 2 kilometers?

Checkpoint

In this investigation, you explored some of the features of driver/follower mechanisms.

ⓐ What is the significance of the transmission factor in the design of rotating objects that are connected?

ⓑ How can you use information about the radii of two connected pulleys, spools, or sprockets to determine the transmission factor?

ⓒ Describe the similarities and differences for two belt-drive systems that have transmission factors of $\frac{2}{3}$ and $-\frac{2}{3}$.

ⓓ If you know how fast a pulley is turning, how can you determine how far a point on its circumference travels in a given amount of time?

Be prepared to share your descriptions and thinking with the entire class.

As you have seen, the transmission factor for rotating circular objects is positive when the two circular objects turn in the same direction. When they turn in opposite directions, the transmission factor is expressed as a negative value. Using negative numbers to indicate the direction opposite of an accepted standard direction is common in mathematics and science.

On Your Own

Wanda's mountain bike has 21 speeds. To get started, she shifts gears so the chain connects the 42-tooth crankset with a 28-tooth rear-wheel sprocket.

a. What is the transmission factor for this system?

b. If Wanda pedals at 40 rpm, at what rate do the rear sprocket and wheel turn?

c. If Wanda pedals at 40 rpm, how far will she travel in one minute if her bike has tires with 64-cm diameters?

The rate (usually in revolutions per minute) at which an object such as a pulley, sprocket, or drive shaft turns is called its **angular velocity**. When Wanda pedaled her bike at 80 rpm, the angular velocity was 80 revolutions per minute. Since one complete revolution turns through 360 degrees, an angular velocity of 80 rpm is also 80 · 360° per minute or 28,800° per minute. The distance that a point on a revolving circle moves in a unit of time is called its **linear velocity**.

In Activity 6 Part c, you found the linear velocity in cm/min of a point on a bike's tire. In Activities 7 and 8, you will explore more fully the idea of angular and linear velocity and how they are affected in systems in which one sprocket or pulley drives another.

7. The Ford Explorer has a tachometer on its instrument panel that gives revolutions per minute (rpm) of the engine.

  a. When the vehicle is driven steadily at about 60 mph in overdrive, the tachometer reads about 2,100 rpm. What is the equivalent angular velocity in degrees per minute?

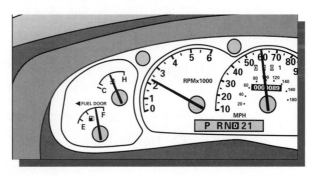

  b. The idle speed of the engine is about 360,000° per minute. What is the reading (in rpm) of the tachometer?

  c. Complete each statement.

  ■ 1 rpm = _____ degrees per minute

  ■ 1 degree per minute = _____ rpm

8. A new 21-speed mountain bike has a crankset of 3 sprockets with 48, 40, and 30 teeth; the bike has 7 sprockets for the rear wheel with 30, 27, 24, 21, 18, 15, and 12 teeth.

a. What is the largest transmission factor available for this bicycle? What is the smallest?

b. Suppose in cross-country riding you can maintain an angular velocity of 70 rpm for the crankset. How fast can you make the rear wheel turn?

c. The radius of a mountain bike tire is about 34 cm. What is the linear velocity of the rear wheel when the crankset turns at 70 rpm and the transmission factor is greatest? When the crankset turns at 70 rpm and the transmission factor is least?

9. The crankshaft of a particular automobile engine has an angular velocity of 1,500 rpm at 30 mph. The crankshaft pulley has a diameter of 10 cm, and it is attached to an air conditioner compressor pulley with a 7-cm diameter and an alternator pulley with a 5-cm diameter.

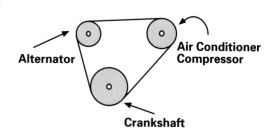

a. At what angular velocities do the compressor and alternator turn?

b. The three pulleys are connected by a 60-cm belt. At a crankshaft rate of 1,500 rpm, how many times will the belt revolve through its 60-cm length in one minute?

c. Most belts do not show significant wear until each point of the belt has traveled about 20,000 kilometers. How long can the engine run at 1,500 rpm before the fan belt typically would show wear?

Checkpoint

Summarize your findings about angular and linear velocity.

a How is *degrees per unit of time* related to *revolutions per unit of time*?

b In what units may the angular velocity and the linear velocity of a point in circular motion be measured? Give a rationale for each unit chosen.

c Explain the relations among the radii, the circumferences, and the transmission factor of two connected circular pulleys.

Be prepared to explain your responses to the whole class.

Although they are no longer used in the United States, foot-operated sewing machines are used in developing countries where electricity either is not available or is too expensive. The foot-operated sewing machine shown below has a driver pulley which is 30 cm in diameter and is attached by a belt to a sewing pulley with a 5-cm diameter. The sewing machine makes one stitch for each revolution of its sewing pulley.

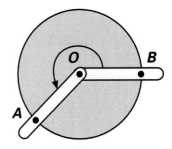

a. The driver pulley is turned easily at one revolution per second.
  - ■ How many stitches does the machine make in 10 minutes?
  - ■ How far would the belt travel in 10 minutes?

b. If the foot pedal (driver) was replaced by an electric motor with a pulley of diameter 1.5 cm, how fast would it need to turn to duplicate the same sewing rate? Express your answer in two ways, revolutions/second and degrees/second.

INVESTIGATION 2 Radian Measure

Angular velocity can be modeled with two strips attached to a cardboard disk as shown below. One strip (in this case, strip *OB*) is fixed like the frame in a linkage and the other strip can turn about the center of the circle. As strip *OA* rotates about *O* in a counterclockwise direction, the angle *AOB* sweeps through measures from 0° to 360°.

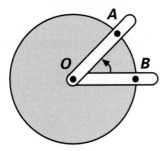

1. Imagine the starting position for strip *OA* has point *A* coinciding with point *B* in the diagram below. Then imagine that strip *OA* rotates through the angles indicated in Parts a–f. Describe the location of point *A* at the end of each rotation.

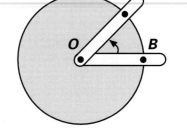

   a. 1 revolution

   b. 360°

   c. $\frac{1}{2}$ revolution

   d. 180°

   e. $2\frac{1}{2}$ revolutions

   f. 45°

2. Now think about the relationship between circular revolutions and degrees.

   a. How many degrees are swept through when strip *OA* rotates $\frac{3}{4}$ revolution? $1\frac{1}{2}$ revolutions? *n* revolutions?

   b. How many revolutions are swept through when strip *OA* rotates 60°? 480°? *n*°?

   c. Complete the following:

      1 revolution = _____ degrees

      1 degree = _____ revolution

Revolutions and degrees are two units by which you can measure ∠*AOB*. In the next activity, you will investigate another unit of angle measure.

3. Working in a group of at least four students, conduct the following experiment. Share the workload among your group members.

   a. With a compass, draw 8 circles with radii varying from 3 cm to 10 cm in increments of 1 cm. For each circle, let *O* be the center of the circle and let *B* be a point on the circle. Use a pipe cleaner, dental floss, or string to locate a new point *A* on the circle so that the length of the arc *AB* is equal to the radius of the circle. Draw ∠*AOB* and measure it, using degrees as your unit. For each circle, record both its radius and the measure of ∠*AOB*. (A shorter way of writing "the measure of ∠*AOB*" is **m∠*AOB*.**)

   b. Examine the numerical data. Describe any patterns you observe. Make and describe the scatterplot of (*radius*, *measure of* ∠*AOB*) data.

   c. What size angle would be determined by a 13-cm arc in a circle with a radius of 13 cm? Explain your thinking.

   d. Suppose a circle with center *O* has radius *r*. What will the measure of ∠*AOB* be if the arc *AB* is *r* units long?

A **radian** is the measure of an angle determined by joining the center of a circle to the endpoints of an arc equal in length to the radius of the circle. (See the diagram below.)

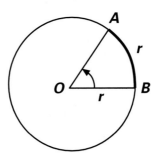

length arc *AB* = *r* = radius
m∠*AOB* = 1 radian

4. Now think about how you could measure angles in radians.

   a. Is 1 radian larger or smaller than 1 revolution? Than 1 degree?

   b. Draw a sketch of a circle with center *O*, radius 8 cm, and a *central angle AOB* which intercepts an arc *AB* of length 16 cm. What is the radian measure of ∠*AOB*? Explain your reasoning.

   c. Draw a sketch of a circle with center *O*, radius 4 inches, and central angle *AOB* which intercepts an arc *AB* of length 22 inches. What is the radian measure of ∠*AOB*? Explain your reasoning.

   d. Reproduced below is the rotating strip model from the beginning of this investigation.

   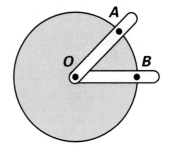

   - Imagine strip *OA* making $\frac{1}{2}$ revolution from a starting point coinciding with strip *OB*. What would be the length of the *intercepted arc AB*? What would be the radian measure of the central angle *AOB*?

   - Imagine strip *OA* making a complete revolution. What would be the radian measure of the angle it sweeps?

   e. Explain why an angle with measure 2π radians is congruent to an angle of 360°.

The degree is the unit of angle measure used in some practical applications, such as surveying, navigation, and industrial design. In scientific work and in advanced work in mathematics, it is usually more convenient to use radian measure. In the next two activities, you will explore how to convert between degree and radian measures.

5. Analyze how Anna and Steve used the fact that 2π radians = 360° to find the radian measure equivalent to 120°.

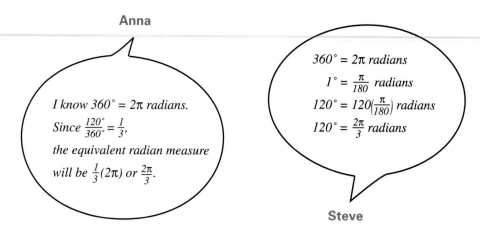

Anna

I know 360° = 2π radians.
Since $\frac{120°}{360°} = \frac{1}{3}$,
the equivalent radian measure
will be $\frac{1}{3}(2\pi)$ or $\frac{2\pi}{3}$.

360° = 2π radians
1° = $\frac{\pi}{180}$ radians
120° = 120$\left(\frac{\pi}{180}\right)$ radians
120° = $\frac{2\pi}{3}$ radians

Steve

   a. How do you think Anna would find the radian measure equivalent to 30°?

   b. How do you think Steve would find the radian measure equivalent to 30°?

   c. Which method, Anna's or Steve's, do you find easiest to use?

   d. How do you think Anna would reason to find the degree measure equivalent to $\frac{\pi}{3}$ radians?

   e. How do you think Steve would reason to find the degree measure equivalent to $\frac{\pi}{3}$ radians?

6. Use any method you prefer to determine the equivalent angle measures in Parts a–c.

   a. Determine the radian measures equivalent to the following degree measures.

     ■ 45°   ■ 30°   ■ 150°   ■ 210°

   b. Determine the degree measures equivalent to the following radian measures.

     ■ $\frac{\pi}{6}$ ■ $\frac{\pi}{3}$ ■ $\frac{3\pi}{2}$ ■ $\frac{11\pi}{6}$

   c. Complete a copy of this table of degree/radian measure equivalents. Save your completed table to use later.

Degree/Radian Equivalents

Degrees	0	30	45	?	90	?	135	150	?	210	?	240	270	300	315	?	360
Radians	?	?	?	$\frac{\pi}{3}$	?	$\frac{2\pi}{3}$	?	?	π	?	$\frac{5\pi}{4}$	?	?	?	?	$\frac{11\pi}{6}$	?

7. Consider once again the information provided by a vehicle's tachometer.

   a. A tachometer of a Ford Explorer reads 2,100 rpm at 60 mph. Find the equivalent angular velocity in degrees per minute and in radians per minute.

   b. The idle speed of a Ford Explorer is 1,000 rpm. Find the angular velocity in radians per minute.

   c. What rpm reading would the tachometer show for an angular velocity of the engine at $6,000\pi$ radians per minute? What would be the degrees per minute equivalent of that angular velocity?

Checkpoint

Angles can be measured in revolutions, degrees, or radians.

   ⓐ In your own words, explain how to draw a 1-radian angle.

   ⓑ Describe how each unit of angle measure below is related to the other two units.

   - Revolutions
   - Radians
   - Degrees

   ⓒ Explain how you would change $m°$ to radians and $\frac{m\pi}{n}$ radians to degrees.

   Be prepared to share your results and thinking with the rest of the class.

▶ On Your Own

The driver pulley of the foot-operated sewing machine in the "On Your Own" on page 419 turns at 4 revolutions per second.

   a. What is its angular velocity in radians per second?

   b. The driver pulley (30-cm diameter) turns a sewing pulley with a 5-cm diameter. What is the angular velocity of the sewing pulley in radians per second?

   c. Suppose a tailor wants the sewing pulley to turn at a rate of 100π radians per second. How fast, in radians per second, must the driver pulley turn to accomplish this rate?

Modeling • Organizing • Reflecting • Extending

Modeling

1. Some bikes used by racing cyclists have 7 sprockets of different sizes connected to the rear wheel and three different-sized sprockets in the crankset attached to the pedals. Typically, the 7 sprockets at the rear have 15, 17, 19, 21, 23, 25, and 27 teeth. The crankset sprockets have 28, 38, and 48 teeth.

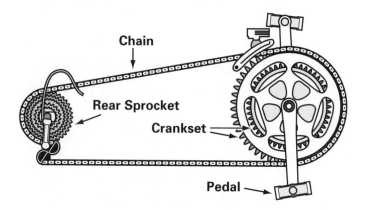

a. What are the two largest crankset-to-rear-sprocket transmission factors? What are the two smallest?

b. One racing bike's wheels have a diameter of about 68 cm. For each of the four arrangements of sprockets in Part a, determine how far a wheel rolls for one revolution of the crankset.

c. In the level-ground touring mode, cyclists may turn the crankset at a rate of 80 times per minute. How fast will the rear wheel turn for the two smallest rear-wheel sprockets connected to the largest crankset sprocket?

d. When cyclists go downhill, sometimes they let the bike "free wheel"—that is, they stop pedaling. Why?

e. Assume the best crankset-to-rear-sprocket advantage and a pedaling rate of 80 rpm. When traveling downhill, what rear-wheel angular velocity would make continued pedaling not useful?

2. The Horse model of the Troy-Bilt Roto Tiller has two forward speeds when the engine drive shaft has angular velocity of 3,000 rpm. The speeds are 0.5 and 1.2 mph.

   a. In feet per minute, how fast does the tiller move at each speed?

   b. The tiller assembly turns at 146 revolutions per minute at both speeds. How many tilling cycles (revolutions) occur in each foot of a garden at each speed?

   c. The wheels on the tiller are 14 inches in diameter. What is the angular velocity of the wheels at each speed?

   d. What is the transmission factor from engine drive shaft to wheels at each speed?

3. Recall that a large driver pulley provides a smaller pulley with greater angular velocity than the driver itself has. For example, a transmission factor of 20 is attained by driving a 4-cm pulley with an 80-cm one. However, physical size constraints often limit the radii of the drivers. To avoid large driver pulleys, two or more pulley-belt systems may be hooked together so that the follower in one directly turns the driver of the second.

   a. Determine by experiment or reasoning the transmission factor from pulley *A* to pulley *B* in the diagram below. Two pulleys on the same shaft, as in the middle of the diagram, turn together and therefore have the same angular velocity.

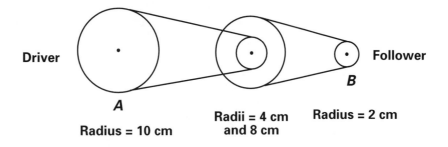

Driver

Follower

B

A

Radius = 10 cm

Radii = 4 cm and 8 cm

Radius = 2 cm

   b. What radius would the driver pulley need to be to produce the same transmission factor found in Part a if the intermediate double pulley were not there? What are the advantages of each setup?

c. Determine the transmission factor from pulley *A* to pulley *B* for each pulley/sprocket setup below. The numbers on the pulleys are their radii in centimeters.

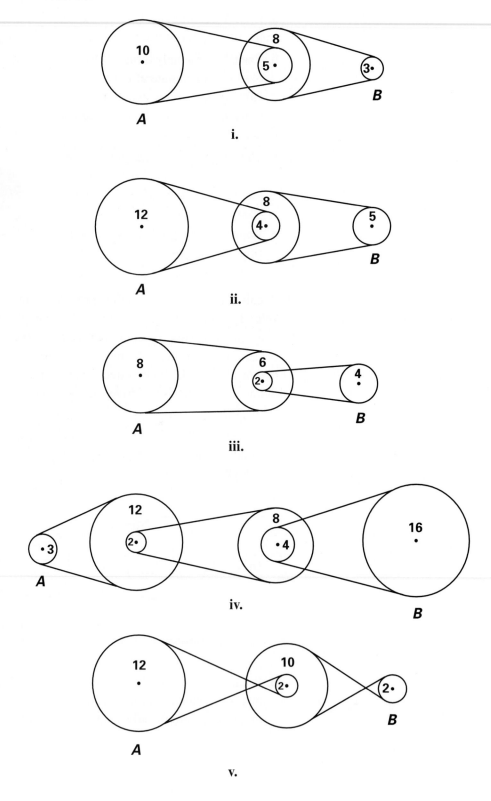

4. In go-carts, the engine-driven sprocket is attached to the rear axle by a belt. These sprockets can have many different numbers of teeth depending upon course demands and safety.

  a. Sketch the situation in which an engine sprocket with a 7-cm diameter drives a rear-axle sprocket with a 10-cm diameter.

  b. Suppose the engine is turning at 1,620 rpm.

   ■ Find the angular velocity of the rear axle.

   ■ Find the go-cart's speed, in km per hour, if the rear wheels have a diameter of 28 cm.

  c. When rounding corners, a speed of 30 km per hour or less is needed to reduce lateral sliding. What engine speed, in rpm, is desirable?

Organizing

1. Some automobile manufacturers are researching an automatic, continuously variable gear based on segments of cones. A simplified model is shown below. It consists of two 10-centimeter segments of right cones. The diameters of the circular ends are given. These partial cones form the basis for a *variable-drive* system, in which either partial cone can be moved laterally (left and right) along a shaft. (Cone-shaped drinking cups can be used to model this situation.)

  a. If the upper shape is the driver, what are the maximum and minimum transmission factors?

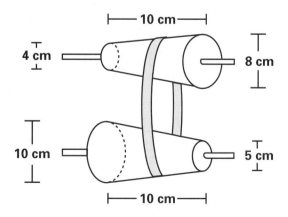

b. Suppose the belt is halfway between the two circular ends of the upper shape. If the lower shape is permitted to move laterally, what range of transmission factors is possible?

c. Describe a position of the belt for which the transmission factor is 1.

2. The transmission factor for pulley (or sprocket) A to pulley (or sprocket) B can be denoted **tf(AB)**. Assume A has radius r_1 and B has radius r_2.

a. What does tf(BA) represent?

b. Express tf(AB) in terms of the radii of A and B. Similarly, express tf(BA).

c. If the circumference of A is C_1 and the circumference of B is C_2, express tf(AB) in terms of C_1 and C_2.

d. Using the formula for the circumference of a circle, rewrite your expression in Part c to one involving only radii (r_1 or r_2). Is this result consistent with Part b? Why or why not?

e. Suppose B turns through an angle of $b°$ whenever A turns through an angle of $a°$. Express tf(AB) in terms of a and b.

3. A line that touches a circle in exactly one point is a **tangent** to the circle. As a belt leaves a pulley, the last point that touches the pulley is a point of tangency. The tangent has a unique relationship to the radius drawn to the tangency point.

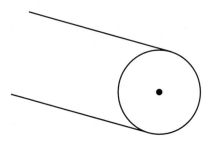

a. Investigate the suggested relationship. First, draw several circles with a compass. For each circle, draw a line that just touches the circle, and measure the angle determined by the radius and the tangent line at that point. What seems to be true in each case? Test your conjecture with another drawing.

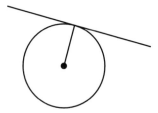

b. In the diagram below, the circle has a radius of 5 cm. The length of $\overline{AB}$ is 15 cm. Segment BC is tangent to the circle at C. How long is $\overline{BC}$?

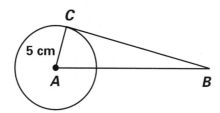

c. Make a copy of the diagram in Part b, above. Then, reflect $\triangle ABC$ across side AB. Let C' be the reflection image of C. Is $\overline{C'B}$ tangent to the circle at C'? Explain your reasoning.

d. Compare the lengths of the two lines from point B, tangent to the circle. What do you observe?

4. A circle of radius r centimeters has a circumference of $2\pi r$ centimeters.

   a. Suppose a point on the circle rotates through an angle of p radians. What is the length of the arc traversed by the point?

   b. Suppose a point rotates at p radians per minute. Find the linear velocity of the point.

   c. Suppose a circle with radius 10 cm has an angular velocity of 80 radians per second. Find the linear velocity of a point on the circle.

   d. Suppose a point on a circle with radius 10 cm has linear velocity of 30π cm per second. Find the angular velocity of the point.

   e. Explain how to convert an angular velocity v (in radians) for a circle of radius r into the linear velocity of a point on the circle.

Reflecting

1. Read the cartoon below. Using mathematical ideas you developed in Investigation 2, write a paragraph explaining to Calvin how two points on a record can move at two different speeds.

2. The *gear* of a bicycle is the product of the transmission factor and the diameter of the rear wheel, commonly 27 inches or 68 cm. What are the gears for the 21-speed mountain bike described in Activity 8 on page 418? How is linear velocity related to "gearing"?

3. The transmission factor of a pulley or sprocket set can be expressed in various ways: in terms of the radii, the diameters, the turning angles, the circumferences, or the numbers of teeth. How are these descriptions related? For you, which representations seem easiest to understand? Why?

4. Refer to the diagram at the right in answering the following questions.

   a. How would you express the radian measure of $\angle AOB$?

   b. In what sense does the radian measure count the number of radii?

   c. Suppose $\angle AOB$ were one third of a complete revolution and r were 4 cm. What would be the radian measure of $\angle AOB$? What would be the degree measure of $\angle AOB$?

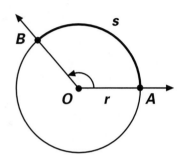

Extending

1. In 1985, John Howard, age 37, of Encinitas, California, rode a specially designed bicycle at the speed of 152.284 mph. Investigate this accomplishment to determine the special conditions of the ride. How did Howard make use of transmission factors?

2. Two pulleys have centers A and B which are p centimeters apart. The radii of circles A and B are r_1 and r_2 respectively, with $r_1 > r_2$.

   a. Find a general expression that can be used to determine the amount of belt in contact with pulley A in a direct-drive pulley system.

   b. Repeat Part a for an opposite-drive system (that is, when the belt is crossed).

3. A record player turntable can rotate at $33\frac{1}{3}$ rpm. On one particular player, the needle arm turns about a pivot point 19 cm from the turntable center. It is 18 cm from pivot to needle. One long-play record has a 15-cm radius with recording beginning at 14.5 cm and ending at 7 cm from the center.

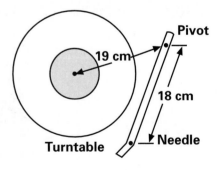

a. If the record is played from start to finish, what path would you see the needle follow?

b. How far does the needle actually move?

c. How far does a point on the outermost (recorded) groove travel in one minute? On the innermost (recorded) groove? What are the linear velocities?

4. A pair of pulleys with 6-cm and 9-cm radii are used to make a belt drive system. They are set 25 cm apart, from center to center.

a. Suppose the pulleys are to turn in the same direction. Find the length of belt needed.

b. Suppose the pulleys are to turn in opposite directions. Find the length of belt needed.

c. Give a mathematical argument supporting the following observation: If two circles have tangents that cross between the circles, then the point of intersection of the two tangents lies on the line containing the centers of the circles. (**Hint:** Think about the symmetry of the situation.)

5. In this task, you will explore a shape that is different in form from a circle but can serve a similar design function.

a. Make a cardboard model of this shape as follows:

- Construct an equilateral triangle of side length 6 cm.
- At each vertex, draw (with a compass) a circle with radius 6 cm, darkening only the small arc between the other two vertices.
- Carefully cut out the shape formed by the darkened arcs.

b. Draw a pair of parallel lines 6 cm apart. Place your model between the lines and roll it along one line. Note any unusual occurrences.

c. Conduct library research on the Wankel engine. How is it related to your model?

d. Investigate the concept of *shapes of constant width*. What are two other examples of such shapes?

INVESTIGATION 3 Modeling Circular Motion

You have seen many examples of circular motion in the previous investigations and in your everyday activities. The motion is so common and used in so many ways that it is important to have mathematical models of circular motion.

The Ferris wheel common to carnivals and fairs is a good example of how circular forms can be used to provide enjoyment to young and old. It provides a good context in which to study the nature of circular motion, especially that of a point on a circle such as a seat on the Ferris wheel.

1. Make a cardboard model of a Ferris wheel by drawing a circle with center C and radius 5 cm. Draw lines to divide circle C into twelve 30° angles, each with vertex C, and then cut out the circle.

   Punch a hole in the center of the circle and, using a paper fastener, attach it to a large piece of paper as shown below. (Be sure your circle can turn freely.) On the paper, draw line PQ as a horizontal axis through center C. Align two of the radii you drew with line PQ, then label the measure of each of the angles in degrees and radians as shown.

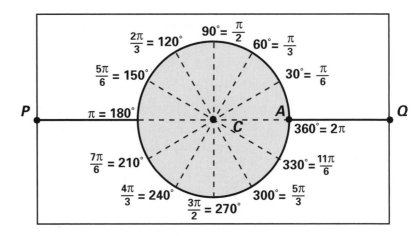

   Mark point A on your circle. Imagine your seat is at A and the Ferris wheel begins to turn counterclockwise.

   a. Describe how your distance from line PQ changes as the wheel turns.

   b. Measure the distance of point A above or below line PQ for each of the angles in your wheel. Represent the distance from point A to line PQ by a negative number when it is below line PQ and by a positive number when it is above line PQ. Enter these *directed distances* in a table similar to the one started below, with 30° increments.

   Note: Leave space for three additional rows to be added later to your table.

Measure of $\angle ACQ$	0°	30°	60°	90°	...	360°
Distance from Point A to Line PQ	0	2.5	?	?	...	?

   c. Make a scatterplot of your (*measure of $\angle ACQ$, distance from point A to line PQ*) data.

   d. The position of point A is a function of the size of $\angle ACQ$. For what size angles is point A above line PQ? How is this shown in the plot?

e. For what angles is point A below line PQ? How is this shown in the plot?

f. For what angles is the directed distance of point A from line PQ a maximum? A minimum? Zero?

g. Use the pattern in the plot to predict the position of point A relative to line PQ when the measure of $\angle ACQ$ is 45°, 195°, 110°, 370°, and 135°.

2. Now draw a vertical line ST on the paper through the center C of the circle.

a. Use your physical model of the Ferris wheel to determine the distances from point A to this vertical line, as the wheel turns counterclockwise. Add a new row to your table labeled "Distance From Point A to Line ST." Record your measures. Use negative numbers when A is to the left of the line.

b. What patterns do you see in the directed distances?

c. Plot your (*measure of* $\angle ACQ$, *distance from point A to line ST*) data using the same scales on the coordinate axes as in Activity 1 Part c.

d. Suppose the Ferris wheel continues to turn. Continue the pattern in the plot to show the distance from point A to the vertical line ST. What additional values should you put on the horizontal axis?

e. Use the plot to predict the position of point A relative to the vertical line ST for angles of 45°, 195°, 110°, 370°, and 135°.

In Activities 1 and 2, you modeled important features of a Ferris wheel using a physical model, and you represented those features with tables and graphs. In the next two activities, you will explore the possibility of describing the patterns of circular motion with equations.

3. The figure below represents your physical model.

a. Using information about the design of your model and a trigonometric ratio, write an expression to calculate the distance from A to PQ after a 30° counterclockwise turn from the horizontal. Do the same for a 60° counterclockwise turn.

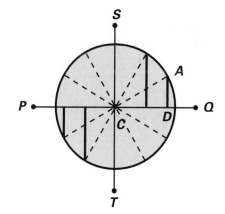

b. Calculate the two distances in Part a. (Make sure your calculator is in degree mode.) Add a row to the bottom of the table you started in Part b of Activity 1 and enter this new information in the table.

c. Determine the directed distances AD for the remaining angles 90°, 120°, 150°, ..., up to 360°. Data lists might be helpful for this. Enter these data in the table.

d. Compare the distances *AD* that you determined by measuring (Activity 1 Part b) and by using the trigonometric ratio.

e. Write an equation that describes the directed distance (*AD*) from point *A* to the horizontal after the wheel turns $x°$ from the horizontal.

f. Produce a graph of your equation. How well does this equation model the data in Activity 1?

4. In this activity, you will investigate a similar relation between the angle of rotation of the Ferris wheel model and the horizontal distance a point is from the line *ST*.

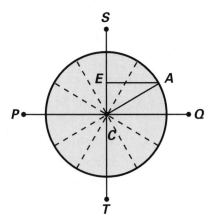

a. Refer to the figure above. Use an appropriate trigonometric ratio to calculate the distance, *AE*, from point *A* to the line *ST* after a counterclockwise 30° rotation. Add another row to the bottom of your table and record this information.

b. Use your calculator to determine the directed horizontal distance *AE* for the remaining angles 60°, 90°, 120°, …, up to 360°. Enter these data in the table.

c. Compare the distances *AE* that you determined by measuring (Activity 2 Part a) and by using the trigonometric ratio.

d. Write an equation that describes the directed distance *AE* that point *A* is from line *ST* after the wheel turns counterclockwise $x°$ from the horizontal.

e. Produce a graph of your equation. How well does this equation model the data in Activity 2?

5. Now, working in pairs and dividing the workload, make scatterplots that correspond to the following when the measure of $\angle ACQ$ is in radians: $\frac{\pi}{6}, \frac{\pi}{3}, \frac{\pi}{2},$ $\frac{2\pi}{3}, \frac{5\pi}{6}, \pi, \frac{7\pi}{6}, \frac{4\pi}{3}, \frac{3\pi}{2}, \frac{5\pi}{3}, \frac{11\pi}{6},$ and 2π.

    i. (*measure of* $\angle ACQ$, *distance from point A to line PQ*) data

    ii. (*measure of* $\angle ACQ$, *distance from point A to line ST*) data

a. How well does the equation you wrote for Activity 3 Part e model the data in Part i above? Make sure your calculator is in radian mode.

b. How well does the equation you wrote for Activity 4 Part d model the data in Part ii above?

Checkpoint

In this investigation, you explored how to model patterns of change associated with circular motion.

a Imagine a point starting in the 3 o'clock position on a circle, as the circle rotates counterclockwise about its center.

■ Describe the pattern of change in the directed distance from the point to the horizontal line through the center of the circle, as the circle makes one complete revolution.

■ Describe the pattern of change in the directed distance from the point to the vertical line through the center of the circle, as the circle makes one complete revolution.

b How are the patterns of change in distances you described in Part a related to the distance patterns obtained as the circle continues to turn?

c Explain how trigonometry can be used to model aspects of circular motion.

d Why is it important to know the mode of a calculator when graphing $y = \sin x$ or $y = \cos x$?

Be prepared to share your descriptions and explanations with the entire class.

▶ On Your Own

At the Chelsea Community Fair, there is a Ferris wheel with a 15-meter radius. Amanuel is on a seat halfway to the highest point of the ride.

a. Make a sketch of Amanuel's position relative to the horizontal and vertical lines through the center of the wheel.

b. Find Amanuel's distance from the lines after the wheel has rotated counterclockwise through an angle of 56°.

c. Write equations that model Amanuel's position relative to the horizontal and vertical lines for any angle $x°$ from his starting point halfway up.

d. Suppose the wheel turns at one revolution every minute. In relation to his starting point, where will Amanuel be at the end of 50 seconds?

INVESTIGATION 4 Patterns of Periodic Change

Many everyday phenomena recur in patterns that are similar year after year. The plot below shows the pattern of change in the number of hours of daylight in Boston over a one-year period, in increments of 10 days. Find the points you think correspond with days in June and days in December.

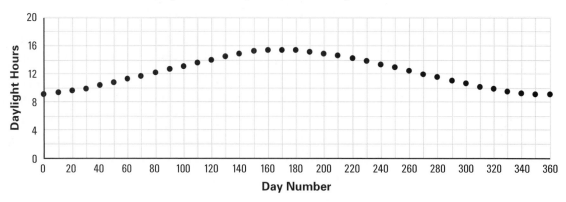

The recurring pattern in length of day is an example of **periodic change**. The complete pattern or *cycle* is seen each year. In fact, it is this cycle that determines the length of a year!

1. Working as a group, identify and describe three or four other real-world situations that exhibit periodic change.

   a. What is the length of the interval over which one complete cycle occurs for your situations?

   b. For each of your situations, make a sketch of the periodic pattern of change on a coordinate system of your choice. Label the axes clearly.

   c. Compare your situations and graphical representations with those of other groups.

Circular motion can generate patterns of periodic change, as you observed in the analysis of the position of a rider on a rotating Ferris wheel in the previous investigation. Now that you have constructed trigonometric models describing these patterns, you are ready to examine, more carefully, properties of the trigonometric models themselves. That is, you will examine the symbolic forms, graphs, and tables of $y = \sin x$, $y = \cos x$, and variations of those equations.

As you work on Activities 2–7, think of x as the measure of an angle with its vertex at the origin and one side on the positive x-axis. Be certain to set the mode of your calculator to "degree" if you want to work in degrees or to "radian" if you want to work in radians.

2. First consider the basic trigonometric models: $y = \sin x$ and $y = \cos x$.

   a. Graph $y = \cos x$ on the interval $0 \leq x \leq 6\pi$. Make a sketch of this graph.

   b. Is this graph periodic? What is the length of one complete cycle? How many cycles are shown in your sketch?

c. What are the maximum and minimum values attained by $y = \cos x$? At what values of x do they occur? At what values of x does the graph intersect the x-axis?

d. Graph $y = \sin x$ on the interval $-2\pi \leq x \leq 4\pi$. Make a sketch of this graph.

e. Is this graph periodic? What is the length of one complete cycle? How many cycles are shown in your sketch?

f. What are the maximum and minimum values attained by $y = \sin x$? At what values of x do they occur? At what values of x does the graph intersect the x-axis?

3. Your analysis of the Ferris wheel model led to two equations, $y = 5 \sin x$ and $y = 5 \cos x$, which described the position of a point on a circle in terms of distance from a horizontal and a vertical line respectively. Graph each of these equations using a graphing calculator or computer software. Use the automatic scaling feature for trigonometric functions (if there is one) to set the initial viewing window. Adjust the range of the y-axis to see a complete graph.

a. Compare the graphs of $y = 5 \cos x$ and $y = \cos x$. Do the same for $y = \sin x$ and $y = 5 \sin x$. How are the graphs in each pair similar, and how are they different?

b. In terms of the Ferris wheel model, why are $\cos x$ and $\sin x$ multiplied by 5 in the rules $y = 5 \cos x$ and $y = 5 \sin x$? What multiplier would you expect if the Ferris wheel had a radius of 8?

c. Now carefully analyze the graph of $y = 5 \cos x$. Use the trace feature to help you. In your analysis, comment on the overall appearance of the graph and each of the following features.

  ■ What are the maximum and minimum y values? What are the corresponding x values?

  ■ What are the possible input values for x?

  ■ For what values of x does the graph cross the x-axis?

  ■ Describe any symmetries you see in the graph.

d. How could you interpret negative input values for x in terms of the Ferris wheel model?

e. How many cycles of the graph are shown in the calculator display?

4. Next, investigate more generally the effect of the multiplier A in $y = A \cos x$.

a. Using your calculator or computer software, graph $y = 3 \cos x$ while the graph of $y = 5 \cos x$ is displayed. What is the effect of changing the 5 to 3?

b. Produce the graph of $y = 7 \cos x$. How is this graph different from the graph of $y = 5 \cos x$? How is it similar? What would you expect for the graph of $y = 4 \cos x$? Why?

c. Now graph $y = \cos x$ and $y = 0.5\cos x$ using your calculator or computer software. Compare the two graphs.

d. How would you modify the graph of $y = \cos x$ to obtain the graph of $y = A \cos x$ when:

- A is a positive number greater than 1?

- A is a positive number between 0 and 1?

e. Investigate the effect of the multiplier A in $y = A \cos x$ when A is a negative number. Explain how it affects the graph for values less than -1 and also for values between -1 and 0.

f. How is the graph of $y = -A \cos x$ related to the graph of $y = A \cos x$?

g. What effect, if any, does the multiplier A in $y = A \cos x$ have on the number of cycles the graph makes for x between $-360°$ and $360°$ or between -2π and 2π? Why does this make sense?

5. Investigate the patterns in graphs of variations of $y = A \sin x$. Again, use the automatic scaling feature (if there is one) of your calculator or computer software to set the initial viewing window. Adjust the range on the y-axis, if necessary, as you complete this activity.

a. Use the trace feature to help analyze the graph of $y = 5 \sin x$. In your analysis, comment on the overall appearance of the graph and each of the following features.

- What are the maximum and minimum y values? What are the corresponding x values?

- What are the possible input values for x?

- For what values of x does the graph cross the x-axis?

- Describe any symmetries you see in the graph.

b. How could you interpret negative input values for x in terms of the Ferris wheel model?

c. How many cycles of the graph are shown in your graph?

d. Investigate the effects of changing the multiplier A in $y = A \sin x$, as was done in Activity 4 for the corresponding cosine rule. Based on your investigation, describe the effect of the multiplier A in $y = A \sin x$ when A is positive and when A is negative.

6. Compare the graphs of $y = \cos x$ and $y = \sin x$ on the interval $-720° \le x \le 720°$ or $-4\pi \le x \le 4\pi$.

a. Are there values of x for which $\sin x = \cos x$? If so, list them.

- Describe how you could find any such points using graphs and using tables.

- How could you interpret these x values in terms of a Ferris wheel model?

b. The graphs of $y = \sin x$ and $y = \cos x$ have a maximum value of 1 and a minimum value of -1.

  ■ In the interval $-720° \le x \le 720°$ or $-4\pi \le x \le 4\pi$, what values of x correspond to the maximum value of $y = \sin x$? To the minimum value of $y = \sin x$?

  ■ In the interval $-720° \le x \le 720°$ or $-4\pi \le x \le 4\pi$, what values of x correspond to the maximum value of $y = \cos x$? To the minimum value of $y = \cos x$?

c. Describe how you could use a transformation of the graph of $y = \sin x$ to obtain the graph of $y = \cos x$.

7. A Ferris wheel operator can manipulate the speed at which the Ferris wheel turns. Consider speeds of 1 rpm (the original speed), 2 rpm, and $\frac{1}{2}$ rpm for a wheel 16 meters in diameter. Let the starting position for the measurement of angles be the 3 o'clock position. Measure the angles counterclockwise from the horizontal line through the center of the wheel.

a. Suppose a seat at the 3 o'clock position rotates through an angle with measure x in a given amount of time when the wheel is turning at 1 rpm. Through what angle will the same seat rotate in the same amount of time when the wheel turns at 2 rpm? At $\frac{1}{2}$ rpm?

b. For the 2 rpm rate, write an equation that models the height of the seat from the horizontal line through the center of the wheel, using the independent variable x as given in Part a. Write a second equation that models the directed distance of the seat from the vertical line through the center. Write corresponding rules for the $\frac{1}{2}$ rpm rate.

c. Graph each of the modeling equations from Part b and compare them to the graphs of $y = \cos x$ and $y = \sin x$.

d. Compare the cycles of the graphs for $\frac{1}{2}$ rpm and 2 rpm with the cycle for 1 rpm. What effect does doubling the angular velocity of the Ferris wheel have on the cycle of the associated sine and cosine models? What effect does halving the angular velocity have on the same models?

e. When working in degrees, $\sin 0° = 0$. At the end of one cycle, $\sin x = 0$ again for what value of x? At the end of one cycle of $y = \sin 2x$, what is the value of x?

f. When working in radians, $\sin 0 = 0$. At the end of one cycle, $\sin x = 0$ again for what value of x? At the end of one cycle of $y = \sin 2x$, what is the value of x?

8. Reproduced below is the graph showing the pattern of change in the number of hours of daylight in Boston over a one-year period. Day 0 is January 1, and the year shown is not a leap year.

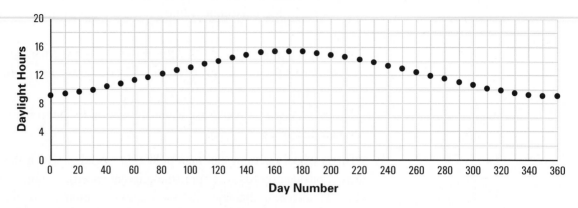

a. If you were to model this situation with a symbolic rule, what would be the independent (input) variable? The dependent variable?

b. Shown below are graphs of $y = \cos x$ and $y = -\cos x$ for $0 \le x \le 2\pi$.

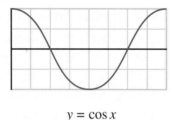 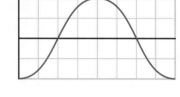

$y = \cos x$ $y = -\cos x$

- Which graph has a shape that approximates the graph of the number of hours of daylight in Boston over a one-year period?

- Describe geometrically how you would transform the chosen graph so that it more closely matches the hours of daylight graph. (Customizing rules for trigonometric models to match patterns of periodic change will be looked at more closely in Course 3.)

- Some calculators and computer software have the capability to find the equation of a best-fitting trigonometric model. The trigonometric regression feature of the TI-83 calculator gives a sine model for these data. An equivalent model that involves the cosine is $y = -3.03 \cos (0.017x) + 12.31$. Does this model look like it would be a good fit for the data? What input values for x make sense in this situation?

```
EDIT CALC TESTS
7↑QuartReg
8:LinReg(a+bx)
9:LnReg
0:ExpReg
A:PwrReg
B:Logistic
C:SinReg
```

c. Recall that the radian measure of an angle whose vertex is the center of a circle is the ratio $\frac{s}{r}$, where s is the length of the arc on the circle that the angle intercepts, and r is the radius of the circle. Since s and r are measured in the same unit, the ratio $\frac{s}{r}$ is a unit-free real number. Thus, radians can be used to model periodic situations in which the independent variable represents a real number, rather than an angle measure. Using radian mode and the modeling equation given in the last section of Part b, estimate:

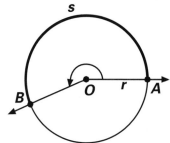

- the hours of daylight in Boston on May 1;

- the days when Boston has the most daylight hours; and

- the days when Boston has more than 12 hours of daylight.

9. Refer back to Activity 1 on page 436, and the situations your group identified as having periodic change.

a. For each of the situations, describe the independent variable and the dependent variable.

b. For each situation, was the independent variable a degree measure or a real number?

Checkpoint

Patterns of periodic change can often be modeled by symbolic rules involving sine or cosine.

ⓐ Describe what is meant by periodic change.

ⓑ Describe the shape of the graphs of $y = \sin x$ and $y = \cos x$.

ⓒ How is the graph of $y = 6 \sin x$ similar to and different from the graph of $y = \sin x$?

ⓓ Compare the graph of $y = -4 \cos x$ to the graph of $y = 4 \cos x$.

ⓔ How is the graph of $y = \sin 2x$ related to the graph of $y = \sin x$?

Be prepared to share your descriptions and comparisons with the class.

A graph that repeats itself over and over again is called a **periodic graph**. The **period** is the length of the smallest interval which contains a cycle of the graph. The period of a graph of $y = \cos Bx$ or $y = \sin Bx$ is determined by the value of B. Often, periodic graphs have a maximum and a minimum in one cycle. When they do, half the difference *maximum value − minimum value* is called the **amplitude**. The amplitude of $y = A \sin x$ or $y = A \cos x$ is $|A|$.

►On Your Own

Portions of periodic graphs are shown below with windows $-720° \leq x \leq 720°$ or $-4\pi \leq x \leq 4\pi$ and $-6 \leq y \leq 6$. Without using a calculator or computer software, match each graph with one of the given rules. In each case, explain the reason for your choice.

- $y = 3\sin x$
- $y = 3\cos x$
- $y = 3 + \sin x$
- $y = -3\sin x$
- $y = -3\cos x$
- $y = \sin 3x$

a.

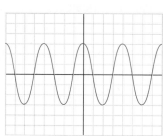

b.

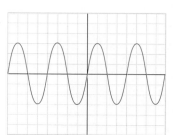

c.

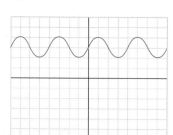

d.

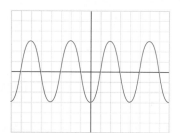

MORE

Modeling • Organizing • Reflecting • Extending

Modeling

1. The center of a Ferris wheel in an amusement park is 7 meters above the ground and the Ferris wheel itself is 12 meters across.

   a. Sketch the Ferris wheel.

   b. Suppose Ashley and her friend enter their seat when it is directly below the center. The wheel takes 20 seconds to make one complete revolution. Sketch a graph by hand showing their height above the ground during 1 minute of this ride.

c. Consider the seat that was at the 3 o'clock position when Ashley entered her seat. Sketch a graph by hand showing the directed distance between that seat and the vertical line through the center of the wheel, as the wheel makes 3 revolutions.

d. Write an equation modeling the periodic motion described in Part c.

2. Refer back to the Ferris wheel ride described in Part b of Modeling Task 1.

a. Make a table of values for the time x and the height y of Ashley's seat above the *ground* during the first minute of the ride. Recall that it takes 20 seconds to make one complete revolution.

b. Use the data pairs in Part a to sketch a graph of the height y as a function of the time x.

c. Is this graph periodic? If so, what is its period?

d. Here are several equations that were predicted to be models of the plot in Part b. Test each to determine if any are good models of these data.

- $y = 6 \sin 18x + 7$ where 18 is angular velocity in degrees per second, x is time in seconds, and y is height in meters.

- $y = -6 \cos 18x$ where 18 is angular velocity in degrees per second, x is time in seconds, and y is height in meters.

- $y = -6 \cos 18x + 7$ where 18 is angular velocity in degrees per second, x is time in seconds, and y is height in meters.

- $y = -6 \cos \frac{\pi x}{10} + 7$ where $\frac{\pi x}{10}$ is angular velocity in radians per second, x is time in seconds, and y is height in meters.

3. An amusement park is planning a new roller coaster. Part of it is to have the shape of a cosine graph. The high and low points of this part differ by 24 meters and cover a horizontal distance of 40 meters from the high point to the low point. The low point is 6 meters below ground in a tunnel. Let y represent the number of meters the track is above or below ground, and let x represent the number of meters the track is horizontally from the high point.

a. Sketch this situation.

b. The equation $y = 12 \cos (4.5x) + 6$ models the situation when x and y are in meters and 4.5 is in degrees per meter. Rewrite $y = 12 \cos (4.5x) + 6$ as an equation for which the coefficient of x is in radians per meter.

Modeling • Organizing • Reflecting • Extending

c. What is the length of a vertical support girder at the high point? At a distance of 6 meters horizontally from the high point? At a distance of 20 meters horizontally from the high point?

d. Measured horizontally, how far from the high point does the roller coaster enter the tunnel?

```
WINDOW
 Tmin=0
 Tmax=360
 Tstep=5
 Xmin=-9
 Xmax=9
 Xscl=1
↓Ymin=-6
```

4. In Investigation 4, you modeled the distance from a point on a circle to a vertical line through the circle's center by the rule $y = 5\cos x$. Similarly, you modeled the distance from the point to the horizontal line through the circle's center with the equation $y = 5\sin x$. Another way to model this situation (which will be developed more fully in Course 4) is to use *parametric equations*. Parametric equations determine a graph using separate equations for the *x*- and *y*-coordinates for each point. Both equations are functions of another *parameter*, usually denoted *T*. Set your calculator or computer software to parametric mode, using degrees. (You may need to consult your manual.)

   Set the viewing window so that *T* varies from 0 to 360 in steps of 5, *x* varies from –9 to 9, and *y* varies from –6 to 6. Then, in the functions list, enter $X_T = 5\cos T$ and $Y_T = 5\sin T$ as the first pair of equations.

   a. Graph and trace values for $T = 30$, 60, 90, and so on.

   b. What physical quantity is represented by X_T? By Y_T?

   c. What measurement does *T* represent? What are the units of *T*?

   d. How much does *T* change in one revolution?

   e. If one revolution of the circle takes 20 seconds, what is the angular velocity of a point on the circle in units of *T* per second?

   f. What is the linear velocity of a point in meters per second if the circle's radius is 5 centimeters?

   g. What equations would model circular motion with a radius of 4 centimeters? Enter them as the second pair of parametric equations in your functions list.

   h. Graph at the same time the equations for circles with radii of 4 and 5 centimeters. (On some calculators, you need to set the mode for simultaneous graphs.) Watch carefully. If the circles both turn at the same angular velocity, on which circle does a point have the greater linear velocity? What is the linear velocity of the slower point?

444 UNIT 6 • GEOMETRIC FORM AND ITS FUNCTION

Organizing

1. Draw a circle with radius 1 unit, centered at the origin O of the coordinate plane. Choose any point $A(x, y)$ on that circle.

   a. Express the coordinates of A in terms of the sine and cosine of an angle at the center of the circle.

   b. Write an equation for the line containing the origin and point A. What is its slope?

   c. Find the tangent of the angle formed by line OA and the positive x-axis (measure counterclockwise).

   d. Compare your results in Parts b and c. What conjecture can you make?

2. In Investigation 4, you used trigonometric models of the form $y = A \sin x$ and $y = A \cos x$ to describe patterns of change associated with a rotating Ferris wheel.

   a. What do tables and graphs of models of the form $y = A \sin x$ have in common?

   b. Use the ideas of geometric transformations to describe how:

   - The graph of $y = -\cos x$ is related to the graph of $y = \cos x$.

   - The graph of $y = A \sin x$, where $A > 0$, is related to the graph of $y = \sin x$.

   - The graph of $y = A \cos x$, where $A < 0$, is related to the graph of $y = \cos x$.

   c. How is the graph of $y = \sin 2x$ related to the graph of $y = \sin x$? How is the graph of $y = \cos 10x$ related to the graph of $y = \cos x$?

   d. In general, when comparing the graphs of $y = \sin x$ and $y = \sin Bx$ or the graphs of $y = \cos x$ and $y = \cos Bx$, what seems to be the effect of the value of B when $B > 0$?

3. The trigonometric ratios (sine, cosine, and tangent) were defined in Lesson 2 for the case of acute angles. In this task, you will explore how the definitions can be extended to angles whose measures are greater than $90°$ or $\frac{\pi}{2}$.

   a. If one side of the angle lies along the positive x-axis, describe in which quadrant the other side of the angle lies for each range of values.

   - $0° < x < 90°$ - $180° < x < 270°$
   - $90° < x < 180°$ - $270° < x < 360°$

   b. For each of the angle ranges in Part a, draw a representative angle. Show how a right triangle can be drawn for each such angle so that the sides of the triangle give the expected trigonometric ratios. What modification in your usual idea of distance needs to be made in order for the ratios to agree in sign (positive or negative) with the calculated values?

   c. Express each range of values from Part a in radian measure.

4. Reproduced below are some of the strip patterns from the "Patterns in Space and Visualization" unit in Course 1.

   a. How are the graphs of periodic models related to these strip patterns?

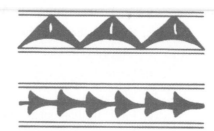

   b. What is the shortest distance you could translate the graph of $y = \sin x$ so that it would map onto itself?

   c. What symmetries, other than translational symmetry, do the graphs of $y = \sin x$ and $y = \cos x$ have? Describe any lines and any centers and angles of rotation involved in each.

5. Draw a circle with center at the origin O and radius 4.

   a. Let $A(x, y)$ be a point on the circle. Consider the angle formed by $\overline{OA}$ and the positive x-axis ($\angle O$). Express the coordinates x and y in terms of the sine and cosine of this angle.

   b. Use the distance formula to express the distance from A to O in terms of the sine and cosine of the angle at the origin. What can you conclude about $(\sin O)^2 + (\cos O)^2$? Do you think this conclusion applies to any angle? Provide evidence to support your answer.

Reflecting

1. Think about the graphs of $y = \sin x$ and $y = \cos x$.

   a. Describe a method you could use to make a quick sketch of each of these graphs.

   b. How is the graph of $y = \cos x$ related to the graph of $y = \sin x$?

2. Karen wishes to evaluate $\sin 25°$ using her calculator. She presses [SIN] 25 and [ENTER]. The calculator displays the screen at the right. What is wrong? What indicates an error?

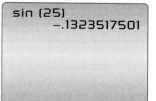

sin (25)
    -.1323517501

3. In *Contemporary Mathematics in Context*, you have studied various patterns of change and how those patterns could be modeled using linear, exponential, power, and trigonometric equations. It is helpful to think about each "family" of algebraic models in terms of basic symbolic rules and their corresponding graphs.

   a. Make and label a general sketch of the graph of each equation below. In each case, $a > 1$.

   - $y = ax$
   - $y = a^x$
   - $y = ax^2$
   - $y = ax^3$

   - $y = \dfrac{a}{x}$
   - $y = \dfrac{a}{x^2}$
   - $y = a \sin x$
   - $y = a \cos x$

   b. Describe one characteristic of the graph of each equation that sets it apart from the graphs of the other equations.

   c. Without sketching, explain how the graphs you drew in Part a would be different if $0 < a < 1$. If $a < 0$.

4. Look back at the graph on page 440 of daylight hours for Boston over a one-year period. How would this graph be different if the number of daylight hours was obtained in Fairbanks, Alaska? At the equator?

Extending

1. Consider a circle of radius 6 with center at the origin.

   a. Lines with slopes 1 and –1 each intersect the circle in two points. Find the coordinates of these points.

   b. If you know one of the points of intersection in Part a, how can you use the symmetry of the circle to determine the remaining three?

   c. The line $y = \dfrac{\sqrt{3}}{3}x$ intersects the circle in two points. Find the coordinates of those points. Find the coordinates of the points on the circle that are symmetric to these points with respect to the *x*-axis. Then find the equation of the line containing these new points.

   d. What angle does $y = \dfrac{\sqrt{3}}{3}x$ make with the *x*-axis?

2. Investigate the effect that adding a constant to x has on the graph of $y = \sin x$ where the input variable is measured in degrees.

   a. Start by graphing $y = \sin x$ and $y = \sin(x + 180)$. How do the two graphs seem to be related?

   b. Experiment by adding different constants to x and then comparing graphs.

   c. Describe the relation between the graphs of $y = \sin x$ and $y = \sin(x + b)$, where b is a constant.

3. Like the various situations you've seen in this lesson, electricity and sound are periodic phenomena that can be modeled by trigonometric equations. A steady electrical current can be created by rotating a rod with copper wire coiled around it through a magnetic field at a constant angular velocity.

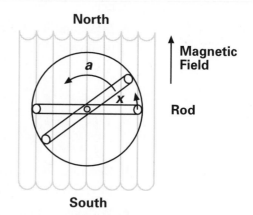

The electricity E created is measured in volts. It turns out that E is a function of the distance from the tip of the rotating rod to the horizontal. The equation is $E = E_0 \sin ax$, where a is the angular velocity in revolutions per second and $x \geq 0$ is time in seconds. For this task, assume $E_0 = 100$ volts and that the angular velocity is 60 revolutions per second.

a. Write an equation for the voltage E in which the angular velocity a is expressed in degrees per second.

b. Graph the equation over the interval from 0 to 2 seconds. What viewing window did you use? Do you think a different interval would give more or better information? Explain.

c. Try graphing the equation for time intervals from 0 to:

- 1 second
- 0.5 second
- 0.1 second
- 0.05 second

Do any of these intervals give a clearer picture of the graph? If so, which ones?

d. Describe how many cycles of the graph you see in Part c when the interval is $0 \leq x \leq 0.1$. Why are there this many cycles?

e. Over what interval would you graph the equation to show just one cycle of the graph? Test your conjecture by graphing the equation over your interval. What interval would you use to show 10 cycles?

f. How can you use 60 revolutions per second to determine the period or length of one cycle?

4. Simple sounds, such as those you can view on an oscilloscope, are modeled by trigonometric equations of the form $y = A \sin Bx$ or $y = A \cos Bx$, where B is expressed in radians per second and x is time in seconds. The amplitude A represents loudness. A good example of a simple sound is the sound produced by a piano tuning fork.

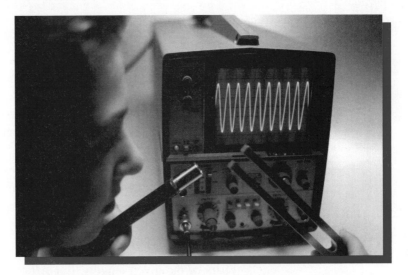

a. The middle C tuning fork oscillates at 264 cycles per second. This is called its frequency. How many radians per second is this?

b. Write a modeling equation for middle C sound when A is 1.

c. What is the period of your model? How is it related to the frequency of 264 cycles per second?

d. Graph your equation. What window should you use to display the graph well?

e. Show four cycles of your model on the graphics screen. If you use 0 for the minimum x value, what should you use for the maximum value?

f. The C note two octaves above middle C has a frequency of 1,056 cycles per second. Model this sound with an equation. What is the period?

g. The C note that is one octave below middle C has a frequency of 132 cycles per second. Model this sound with an equation. Graph your modeling equation.

h. Graph your modeling equations for Parts b and g in the same viewing window. What do you think is the frequency of the C note one octave above middle C?

Looking Back

As you complete this unit, you may be wondering whether you have been studying mathematics or principles of engineering. That should not be surprising. In many applications of geometric shapes and principles, form and function are intertwined. You have seen that linkages and circular shapes, while simple in design, are useful in many different situations. Their usefulness is directly tied to their geometric properties. This final lesson will help you review and apply the major ideas in this unit.

1. Examine this 4-bar linkage with bars of the given lengths. Bar *AB* is the fixed frame.

   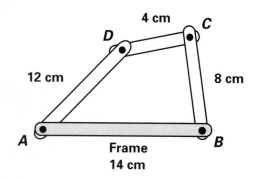

   a. Will either bar *AD* or bar *BC* make complete rotations about point *A* or point *B*? Explain.

   b. Using a scale drawing or a model, determine the largest angle that bar *AD* can make with bar *AB* (∠*DAB*).

   c. Find the smallest measure of ∠*DAB*.

   d. What is the angle through which bar *AD* rotates as bar *BC* is moved?

2. Indy-cars are special race cars that use a parallelogram linkage in the front suspension. The wheel is attached to the coupler, and the frame is attached to the car so that the frame is perpendicular to the ground. The linkage can pivot freely at points *A*, *B*, *C*, and *D*. When the car is not moving, the full width of the tire is in contact with the pavement.

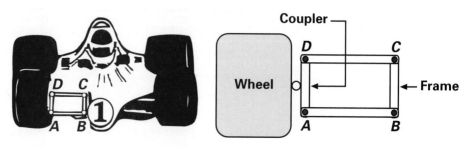

a. Describe the motion of the tire as the entire suspension linkage moves up and down.

b. If the tire is parallel to the coupler when the car is not moving, how is it related to the coupler when the car is in motion?

c. At the speeds that Indy-cars travel, it is important that they "grip" the road by having as great tire-to-pavement contact as possible. How does this suspension linkage design accomplish this goal?

3. Early spinning wheels were run by a *foot treadle* and a *flywheel*. The operator pushed the flywheel to start it turning, then kept it spinning by pressing the foot treadle at appropriate times. The spinning wheel on the left below has a flywheel with radius 15 inches. It is attached to a 2-inch radius pulley, called a *bobbin*, on which the yarn accumulates.

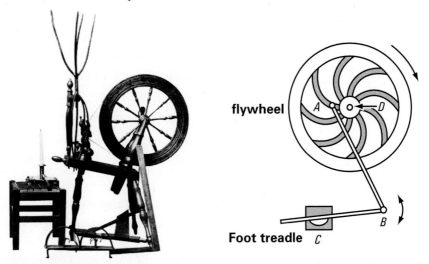

a. What is the transmission factor from the flywheel to the bobbin? An accomplished artisan can make the flywheel turn at 150 rpm. How fast will the bobbin rotate?

b. The diagram on the right above illustrates how the foot treadle turns the flywheel. Identify the 4-bar linkage that makes the spinning wheel work. What is the driver of this linkage? The follower? The frame?

c. If the length of $\overline{AD}$ is 3 inches, describe what is modeled by the equation $y = 3 \sin x$ where x is the degree measure of the angle that $\overline{AD}$ makes with the horizontal through point D.

d. If the center of the flywheel is 30 inches above the floor, how should the equation in Part c be modified to give the height of point A from the floor?

4. Depth sounders such as the one shown on the next page are used on pleasure and fishing boats. The sounder *transducer* sends out a signal that returns information to the boat. The screen will display the shape of the bottom of the lake or river, the depth of the water, and any fish that happen to be in the path of the signal. Sport fishers on inland, fresh-water lakes use the depth sounder to examine the characteristics of the bottom and to locate fish.

The transducer on one depth sounder sends out two signals. Each is in the shape of a right circular cone. One has a vertex angle of 16°, the other is 53°.

a. Make a sketch showing the depth sounder cones for both angles.

■ The promotional material for this depth sounder claims the 16° angle signal identifies fish directly under a 14-foot boat that is about 5 feet wide. Is this a reasonable claim if fish are marked at a depth of 20 feet? Explain.

■ How large is the radius of the 16° angle-signal cone when fish are marked at 10 feet? At 30 feet? At 50 feet?

b. The narrow-angle and wide-angle signals from the transducer are sent out at the same time. Fish in the wide-angle signal are shown differently on the display screen than those in the narrow-angle signal. Suppose a fish is spotted at 30 feet in the wide-angle cone. Other than the fact that the fish is at 30 feet, what else can you say about the location of the fish?

■ What is the radius of the wide-angle cone when fish are spotted at 10 feet? At 20 feet? At 50 feet?

■ How are the radii of the two signal cones related at the same depth?

■ When the radius of the narrow cone is about 5 feet, what is the radius of the wide-angle cone at the same depth? What is that approximate depth?

c. How may trigonometric models be used to determine the radius of the circular field of the 16° signal at a depth of 100 feet? Of the 53° signal at a depth of 100 feet?

5. Suppose that you are trying to model the motion of a clock pendulum that moves as far as 5 inches to the right of vertical and swings with a period of 2 seconds.

a. Experiment with variations of the rule $y = \cos x$ to find:

■ a modeling equation whose values range from –5 to 5 and has a period of 2π;

■ a modeling equation that has a period of 2 and whose values range from –1 to 1; and

■ a modeling equation that has a period of 2 and whose values range from –5 to 5.

b. How are the numbers in the final symbolic rule from Part a related to the motion of the pendulum you are modeling?

c. Graph the equation that models the motion of the clock pendulum. Identify the coordinates of the *x*-intercepts and minimum and maximum points of the graph.

Checkpoint

In this unit, you have studied how mechanisms work and how their function is related directly to the form or shape of the mechanism. You also investigated how some patterns of periodic change could be modeled.

ⓐ What characteristics of a parallelogram make the shape widely useful as a linkage?

ⓑ Two plane shapes are similar with a scale factor *k*. How are the lengths of corresponding segments related? How are measures of corresponding angles related? How are areas of corresponding regions related?

ⓒ Define the sine, cosine, and tangent of an acute angle of a right triangle. How can these ratios be used to determine lengths that cannot be measured directly? How can these ratios be used to determine angle measures that cannot be measured directly?

ⓓ How does the angular velocity of one rotating pulley or gear affect the angular velocity of a second pulley or gear connected to it by a belt, if the radii are r_1 and r_2 and the following is true:

   ■ $r_1 = r_2$ ■ $r_1 > r_2$ ■ $r_1 < r_2$

ⓔ Although often measured in degrees, angles also can be measured in radians. Describe what it means to say an angle has radian measure 2. Describe another use for radians.

ⓕ Why are variations of trigonometric rules such as $y = a \sin bx$ or $y = a \cos bx$ often used to model periodic change? What does the value of *a* tell you about the situation being modeled? What does the value of *b* tell you?

Be prepared to explain your responses to the entire class.

▶ On Your Own

Write, in outline form, a summary of the important mathematical concepts and methods developed in this unit. Organize your summary so that it can be used as a quick reference in future units and courses.

Patterns in Chance

Waiting Times

Monopoly® is a board game that can be played by several players. Movement around the board is determined by rolling a pair of dice. Winning is based on a combination of chance and a sense for making smart real estate deals. While playing Monopoly, Anita draws the card shown below. She must go directly to the "jail" space on the board.

Anita may get out of jail by rolling doubles with a pair of dice on one of her next three turns. Doubles means that both dice show the same number on the top. If she does not roll doubles on any of the three turns, Anita must pay a $50 fine to get out of jail. Anita takes her first turn and doesn't roll doubles. On her second turn, she doesn't roll doubles again. On her third and final try, Anita doesn't roll doubles yet again. She grudgingly pays the $50 to get out of jail. Anita is feeling very unlucky.

Think About This Situation

Anita's situation suggests several questions.

ⓐ How likely is it that a Monopoly player who is sent to jail (and doesn't have a "Get Out of Jail Free" card) will have to pay $50 to leave? As a class, think of as many ways to find the answer to this question as you can.

ⓑ In games and in real life, people are occasionally in the position of waiting for an *event* to happen. In some cases, the event becomes more and more likely to happen with each opportunity. In some cases, the event becomes less and less likely to happen with each opportunity. Does the chance of rolling doubles change each time Anita rolls the dice?

ⓒ On average, how many rolls do you think it takes to roll doubles? Do you think Anita should feel unlucky? Explain your reasoning.

INVESTIGATION 1 ▶ Waiting for Doubles

In this investigation, you will explore several aspects of Anita's situation. For this investigation, we will change the rules of Monopoly so that a player must stay in jail until he or she rolls doubles. A player cannot pay $50 to get out of jail in this version of the game, and there is no "Get Out of Jail Free" card.

1. Now suppose you are playing Monopoly under this new rule and have just been sent to jail. Take your first turn and roll a pair of dice. Did you roll doubles and get out of jail? If so, stop. If not, roll again. Did you roll doubles and get out of jail on your second turn? If so, stop. If not, roll again. Did you roll doubles and get out of jail on your third turn? If so, stop. If not, keep rolling until you get doubles.

   a. Copy the frequency table below and put a tally mark in the frequency column next to the event that happened to you. Add rows as needed.

Rolling Doubles

Event	Number of Rolls	Frequency
Rolled doubles on first try	1	
Rolled doubles on second try	2	
Rolled doubles on third try	3	
Rolled doubles on fourth try	4	
⋮	⋮	
Total		100

b. With other members of your class, perform this experiment a total of 100 times. Record the results in your frequency table.

c. Do the events in the frequency table appear to be **equally likely**? That is, does each of the events have the same chance of happening?

d. Use your frequency table to estimate the probability that Anita will have to pay $50, or use a "Get Out of Jail Free" card, to get out of jail when playing a standard version of Monopoly. Compare this estimate with your original estimate in Part a of the "Think About This Situation" on page 457.

e. Make a histogram of the data in your frequency table. Describe the shape of this histogram.

f. Explain why the frequencies in your table are decreasing even though the probability of rolling doubles on each attempt does not change.

2. Later in this unit, you will analyze Anita's situation theoretically. That is, you will use mathematical principles to find the probability she has to pay $50. As a first step, in this activity you will explore how to find the probability of various events when two dice are rolled.

a. Suppose a red die and a green die are rolled at the same time. Make a copy of the matrix-like chart below.

Rolling Two Dice

Number on Red Die \ Number on Green Die	1	2	3	4	5	6
1	1, 1					
2						
3		3, 2				
4					4, 5	
5						
6						

- What does the entry "3, 2" mean?
- Complete the chart, showing all possible outcomes when the two dice are rolled.
- How many outcomes are possible?
- Are these outcomes equally likely? Why or why not?

b. If two dice are rolled, what is the probability of getting each of the following events?

- Doubles
- A sum of 7
- A sum of 11
- A sum of 7 or a sum of 11
- Either a 2 on one or both dice or a sum of 2

c. Is the probability of rolling doubles the same if both dice are the same color? Explain your reasoning.

d. Suppose that in playing the modified Monopoly game, Anita is still in jail after trying twice to roll doubles. Conchita has just been sent to jail. Does Anita or Conchita have a better chance of rolling doubles on her next turn? Compare your answer with that of other groups. Resolve any differences.

Checkpoint

In this investigation, you explored the waiting time for rolling doubles.

a Suppose you compared your class's histogram of the waiting time for rolling doubles with another class's histogram.

- Explain why the histograms should or should not be exactly the same.
- What characteristics do you think the histograms will have in common?

b If two dice are rolled several times, what is the probability of getting doubles on the first roll? On the fourth roll?

Be prepared to share your ideas with the entire class.

On Your Own

Change the rules of Monopoly so that a player must flip a coin and get heads in order to get out of jail.

a. Is it harder or easier to get out of jail with this new rule instead of by rolling doubles? Explain your reasoning.

b. Play this version 24 times, either with a coin or by simulating the situation. Put your results in a table like the one in Activity 1 of Investigation 1. Then make a histogram of your results.

c. What is your estimate of the probability that a player will get out of jail in three flips or fewer?

INVESTIGATION 2 Independent Trials

In some situations, the probability of getting a particular event is the same on every opportunity (called a **trial**), no matter what happened on previous trials. In these cases, the trials are said to be **independent**. In other situations, the probability of getting a particular event changes depending on what has happened on previous trials. In these cases, the trials are called **dependent**.

1. In the previous investigation, you explored probabilities associated with rolling two dice. Are the rolls of dice independent or dependent? Explain your reasoning.

2. Think about the question of independent or dependent trials as you examine the following games of chance. To play these games, you will need a bag with one red and one yellow marker or token in it. You will also need some extra red and yellow markers. In each game, the goal is to draw a red marker. When that happens, the game stops and the player's score is the number of draws required. Each player starts his or her turn with one red and one yellow marker in the bag.

   a. First, analyze each of these games. Which game would you choose to play if you get a prize for having the smallest score?

   Game A: Draw until you get a red marker. Replace the marker after each draw. In addition, each time you draw a yellow marker, you must add another yellow marker to the bag before drawing again.

   Game B: Draw until you get a red marker. Replace the marker after each draw. In addition, each time you draw a yellow marker, you must add a red marker to the bag before drawing again.

   Game C: Draw until you get a red marker. Replace the marker after each draw, but don't add any other markers to the bag. (There are always the original two markers in the bag.)

   b. Now, play each game ten times and complete a frequency table like the one below, showing the number of draws required until a red marker appears. Share the work by having different members of your group play and record results for different games.

Three Different Games

Number of Draws to Get a Red	Game A	Game B	Game C
1			
2			
3			
4			
⋮			
Total	10	10	10

c. In which game does the probability of drawing a red marker get larger with each trial? Get smaller with each trial?

d. In which game are the trials independent?

e. Do you want to change your answer to Part a? Why or why not?

3. In each of the following situations, imagine you are waiting for an event to occur. In each case, decide if the trials are independent. Give reasons for your answers.

a. The event is getting an ace. A trial is drawing a card from a deck of playing cards.

b. The event is winning a weekly lottery. A trial is buying one lottery ticket each week.

c. Ten students from your class will be selected at random. The name of each student is written on a slip of paper and the slips are placed in a box. One slip of paper is drawn at a time and not replaced before the next name is drawn. The event you are waiting for is to hear your name called. A trial is one name being drawn.

d. The event is getting the prize in a box of cereal that has been mixed well. The prize can be anywhere in the box, not just at the bottom or top. A trial is pouring and eating a bowl of cereal from the box. The trials continue each morning until you get the prize.

e. The event is having a daughter. A trial is having a baby.

Checkpoint

Think about situations involving chance that are different from those in this investigation.

ⓐ Give an example of trials that are dependent.

ⓑ Give an example of trials that are independent.

Be prepared to explain your examples to the class.

▶On Your Own

Check your understanding of independent and dependent trials.

a. The modified version of Monopoly in the "On Your Own" on page 459 required the player to flip a coin and get heads in order to get out of jail. Are the trials independent in that version of Monopoly? Explain.

b. Suppose the event is seeing a species of bird that you have never seen before. A trial is going to the park to watch birds. Are the trials independent? Give a reason for your answer.

INVESTIGATION 3 The Distribution of Waiting Times

According to the company that makes them, "M&M's"® Chocolate Candies are put randomly into bags from a large vat in which all the colors have been mixed. This means we can assume that a bag of these candies is a random sample. It also means that the probability of getting a brown candy is almost exactly the same each time you draw a candy out of any bag.

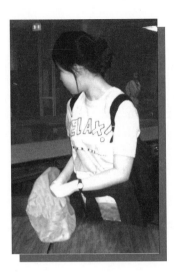

1. Suppose each of five students took "M&M's"® Chocolate Candy one at a time out of a large bag until he or she got a brown one. Marcia's fifth candy was her first brown one. The list below gives the number of draws that each student needed to take to get a brown candy.

   Marcia   5
   Jenny   10
   David   3
   Whitney   1
   Simon   3

   a. What was the average (mean) number of candies that had to be taken out of the bag in order to get a brown one?

   b. The answer to Part a isn't a whole number, so it isn't really possible to make that number of draws. Give an example of another situation in which the average is not one of the possible events.

2. You may remember discovering, in Course 1, an easy method for calculating the mean of a frequency table. That method can be summarized by the following formula.

$$\overline{x} = \frac{\sum xf}{n}$$

Here, x is a data value, f is the frequency of that value, and n is the sum of the frequencies. The $\sum$ stands for "sum." When you use this formula, it is often efficient to add a third column, xf, to the frequency table.

   a. Use this formula to find the mean of the frequency table below.

x	f
0	9
1	5
2	6
3	4

   b. Explain how you could use (or did use) the list features of a graphing calculator or computer software to find the mean of this frequency table.

3. Suppose each student in a group of 1,000 drew "M&M's"® Chocolate Candies one at a time out of individual bags until he or she got a brown one. A trial is taking a candy out of the bag. Each student counted the number of trials until a brown candy was drawn. A typical frequency distribution of the number of draws to get a brown candy appears below. The frequency distribution for this experiment is an example of a **waiting-time** (or **geometric**) **distribution**. (Waiting for doubles when playing Monopoly also has a waiting-time distribution.)

Drawing Candies

Number of Draws to Get First Brown Candy (Event)	Number of Students (Frequency)
1	284
2	197
3	163
4	91
5	88
6	52
7	39
8	30
9	15
10	14
11	13
12	5
13	6
14	1
15	1
16	0
17	0
18	1
Total	1,000

a. How many students got their first brown candy in three draws or fewer?

b. Make an estimate of the probability that the first brown candy will appear on or before the third draw when a person draws candies out of a bag.

c. Make an estimate of the probability that the first brown candy will not appear until the sixth draw or later.

d. Examine the histogram of the distribution shown below.

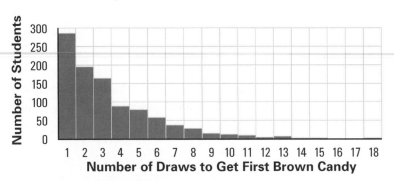

- Describe the histogram's basic shape.
- Estimate the average (mean) number of draws it takes to get the first brown candy by estimating the point at which this histogram "balances."
- Why are the heights of the bars in the histogram decreasing?

e. Now calculate the average number of draws to get the first brown candy. Compare the calculated value with your estimate in Part d.

f. How can you use the table to estimate the percentage of candies that are brown? What is your estimate of this percentage?

g. Estimate the probability of drawing 11 candies or more before getting a brown one.

h. Marty is amazed that it took him so many draws to get his first brown candy. He noticed that 95% of the students got a brown candy before he did. How many candies did Marty draw?

Marty's result fell in the upper 5% of the waiting-time distribution for drawing a brown candy. When they have to wait so much longer than almost everyone else, lots of people begin to feel that something very unusual has happened. Thus, we will define a **rare event** as an event that falls in the upper 5% of a waiting-time distribution.

i. Is having to draw 11 candies before getting a brown one a rare event? What about having to draw 8?

Checkpoint

Look over the situations with waiting-time distributions you have seen so far.

a How are the histograms similar?

b How can you estimate the average waiting time using a histogram? How can you calculate it using a frequency table?

c How can you determine if a specific event in a waiting-time distribution is a rare event?

Be prepared to share your ideas and methods with the entire class.

On Your Own

Cereal manufacturers often place small prizes in their cereal boxes as a marketing scheme. Boxes of one brand of cereal recently contained one of three basketball-player posters: Patrick Ewing, Alonzo Mourning, and Shawn Kemp. Suppose that equal numbers of these posters are placed randomly into the boxes. Patrick would like to get a Patrick Ewing poster.

a. On the average, how many boxes do you think Patrick would have to buy before getting a Patrick Ewing poster?

b. Designing simulations was an important part of the "Simulation Models" unit in Course 1. Design a simulation so that you can better estimate how many boxes of this cereal Patrick might have to buy before he gets a Patrick Ewing poster.

c. Repeat your simulation five times.

d. How many boxes did Patrick have to buy each time? Add your results to the frequency table below so that there is a total of 100 trials.

Getting a Patrick Ewing Poster

Number of Boxes	Frequency	Number of Boxes	Frequency
1	31	7	3
2	20	8	2
3	15	9	1
4	10	10	1
5	6	11	1
6	4	12	1
		Total	

e. Make a histogram of the results in the completed frequency table and describe its basic shape.

f. Use the completed frequency table to estimate how many boxes a person would have to buy to get a Patrick Ewing poster.

g. Estimate the chance that a person would have to buy more than 10 boxes to get a Ewing poster.

h. Is having to buy 6 boxes a rare event?

Modeling

1. Suppose you are trying to draw a heart from a regular deck of 52 cards.

   a. After each draw, you *do not* replace that card before you draw again.

   - What is the smallest number of cards you might have to draw in order to get a heart?

   - What is the largest number of cards you might have to draw in order to get a heart?

   - Are the draws independent? Explain.

   b. After each draw, you *do* replace that card (and reshuffle) before you draw again.

   - What is the smallest number of cards you might have to draw in order to get a heart?

   - What is the largest number of cards you might have to draw in order to get a heart?

   - Are the draws independent? Explain.

   c. Should you replace the card or not if you want to get a heart in the fewest number of draws? Why does this make sense?

2. Describe a simulation using random digits to estimate the number of draws from a regular deck of cards needed to get a heart if the card is replaced after each draw. Repeat your simulation 10 times.

   a. Based on your simulation, what is the average number of draws needed to get a heart?

   b. How would you modify your simulation model if the card is *not* replaced after each draw?

3. Boxes of Kellogg's® Cocoa Krispies® cereal once contained one of four endangered animal stickers: bird of paradise, tiger, African elephant, and crocodile. Suppose that these stickers were placed randomly into the boxes and that there were an equal number of each kind of animal sticker.

   a. Polly likes birds and wanted the bird of paradise sticker. Describe a simulation to estimate the average number of boxes of Kellogg's® Cocoa Krispies® cereal that Polly would have had to buy before she got a bird sticker.

©1993 Kellogg Company

b. Repeat your simulation five times. Add your results to those in the frequency table below so that there is a total of 100 trials. Add additional rows if you need to.

Getting a Bird Sticker

Number of Boxes	Frequency	Number of Boxes	Frequency	Number of Boxes	Frequency
1	19	10	1	19	1
2	14	11	1	20	0
3	15	12	3	21	0
4	13	13	0	22	0
5	11	14	0	23	0
6	6	15	0	24	0
7	4	16	2	25	0
8	2	17	1	26	1
9	1	18	0	**Total**	

c. Make a histogram from the completed frequency table and describe its shape.

d. Use the completed frequency table to estimate the average number of boxes a person would have to buy to get a bird of paradise sticker.

e. Estimate the chance that a person would need to buy more than 10 boxes.

f. Is having to buy 10 boxes a rare event?

4. Backgammon is one of the oldest games in recorded history. It may have originated before 3000 B.C. in Mesopotamia (present-day Iraq). Today, it is played all over the world. In Backgammon, as in Monopoly, the number of spaces you can move a stone is determined by rolling two dice. In the game of Backgammon, if you "hit" another player's single stone exactly, that stone must go back to the beginning and start again.

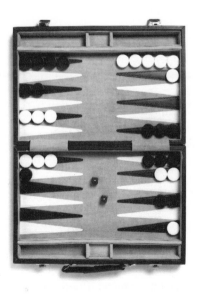

To hit a stone that is three spaces ahead of you, you must roll a three. The three may be on one die or the three may be the sum of both dice. If you roll double 1s, you also can hit the stone because on doubles a player gets to move the numbers that show on the die twice each. So a player who rolls double ones could move 1+1+1, hit the stone, and then move the final 1.

a. What is the probability of being able to hit the stone of a player who is three spaces ahead of you?

b. What is the probability of being able to hit the stone of a player who is five spaces ahead of you?

c. What is the probability of being able to hit the stone of a player who is twelve spaces ahead of you?

Organizing

1. Refer to the frequency table your class prepared for Activity 1, Investigation 1 (page 457).

   a. Make a scatterplot of the (*number of rolls required, frequency*) data.

   b. If *NOW* is the number of people who got doubles on a roll and *NEXT* is the number of people who got doubles on the next roll, write an equation that approximates the relation between *NOW* and *NEXT*. Explain why your equation is reasonable in the context of rolling doubles.

   c. Is a linear model or an exponential model a better fit to the scatterplot? Why?

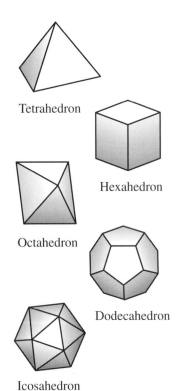

Tetrahedron

Hexahedron

Octahedron

Dodecahedron

Icosahedron

2. Recall that there are five regular polyhedra: tetrahedron (4 faces), hexahedron or cube (6 faces), octahedron (8 faces), dodecahedron (12 faces), and icosahedron (20 faces). Find or imagine pairs of dice in the shape of these polyhedra. Rolling each tetrahedral die generates numbers from 1 to 4, rolling each octahedral die generates numbers from 1 to 8, and so on.

   a. Make a chart like the one in Activity 2 of Investigation 1 (page 458) for a pair of tetrahedral dice. What is the probability of getting doubles?

   b. Repeat Part a in the case of a pair of octahedral dice.

   c. By looking for patterns in your work, find the probability of rolling doubles with the following pairs of dice:

   ■ Dodecahedral dice

   ■ Icosahedral dice

   d. For which pair of dice is the probability of getting doubles the greatest?

   e. If the number of faces on each of a pair of regular polyhedral dice is n, what is the probability of rolling doubles with that pair of dice?

3. Describe how you could use a calculator to simulate rolling an icosahedral die.

4. In the "Network Optimization" unit, you explored ways in which special vertex-edge graphs called trees could be used to model situations. Think of a way to use a tree graph to represent rolling two dice. Illustrate how you could use the tree graph to answer the questions in Activity 2, Part b, from Investigation 1 (page 459).

Reflecting

1. Suppose someone in your class is unsure about whether the probability of rolling doubles with one red die and one green die is the same as if both dice were red. Describe an experiment you could do to convince the person.

2. Take a survey of your friends and family. Tell them about the problem of getting out of jail in Monopoly and then ask them to answer this question:

   A player has tried twice to get out of jail. She has had no luck. Before her third try she says, "I have missed getting out of jail twice, so I'm due for doubles. My chances of rolling doubles are greater this time than on my first two tries." Is she correct?

   Are you surprised at their answers? Why do some people believe that the chances of success increase after there have been several failures?

3. If you have not previously played Monopoly, learn how to play. Make a list of three probability questions that arose during your game. Find the answer to one of your questions by using a simulation.

4. Suppose a friend looks at the table in Activity 1 of Investigation 1 (page 457) and says, "I don't understand why the frequency for 'Rolled doubles on second try' is smaller than 'Rolled doubles on first try.' They should be equal because the probability of rolling doubles is the same, $\frac{1}{6}$, on each try." What would you say to this friend?

Extending

1. In a famous trial in Sweden, a parking officer had noted the position of the valve stems on the tires on one side of a car; upon returning later, the officer noted that the valve stems were still in the same position. The officer noted the position of the valve stems to the nearest "hour." For example, in the following picture, the valve stems are at 3:00 and at 10:00. The officer issued a ticket for overtime parking. However, the owner of the car claimed he had moved the car and returned to the same parking place.

   Design a simulation to estimate the probability that if a car is moved, the valve stems return to the same position they had before the car was moved. You must make an assumption about how tires rotate. Try to find out whether this assumption is true or not before doing your simulation. Do you think the car owner should have been found guilty or innocent of the parking violation? Explain.

2. In the game of Backgammon, if you want to have the best chance of hitting an opponent's stone with a particular stone of your own on the next roll of dice, how many spaces away should your stone be? (The rules of Backgammon are explained in Modeling Task 4 on page 467.)

3. Shown below are labeled nets for special dice. Darnell selects one of the special dice and then Joy selects one of the remaining two.

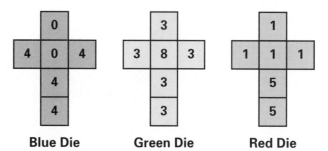

Blue Die   **Green Die**   **Red Die**

Each rolls his or her die. The person with the larger number wins. To help you decide if it is better to use, for example, the blue die or the green die, you might want to complete a table like the following:

Number on Green Die

		3	3	3	3	3	8
0	Green Die Wins						
0	Green Die Wins						
4	Blue Die Wins						
4							
4							
4							

Number on Blue Die

Which die should Darnell choose? Which should Joy choose? What is the surprise here? Can you find a different set of dice that has the same property?

4. Put two red and two yellow markers or tokens in a bag. Without looking, draw one. Replace it with a marker of the other color. Continue this process until all the markers in the bag are the same color.

a. How many draws did it take until all markers in the bag were the same color?

b. Make a frequency table and put your result in the table.

c. Repeat this experiment 19 more times and place your results in the table. (You may want to combine your results with those of several classmates.)

d. Make a histogram from your frequency table.

e. Does this experiment generate a waiting-time distribution? Explain.

f. Estimate the average number of draws needed until all the markers are the same color.

g. Design a simulation that uses a random device instead of a bag of markers.

The Multiplication Rule

Some physical characteristics, such as freckles, eyelash length, and the ability to roll one's tongue up from the sides, are determined in a simple manner by genes inherited from one's parents. Each person has two genes that determine whether or not he or she will have freckles, one inherited from the father and one from the mother. If a child gets a "freckles" gene from either parent or from both parents, the child has freckles. In order not to have freckles, the child must inherit a "no-freckles" gene from both parents. This explains why the gene for freckles is called *dominant* and the gene for no-freckles is called *recessive*. A parent with two freckles genes must pass on a freckles gene to the child; a parent with two no-freckles genes must pass on a no-freckles gene. If a parent has one of each, the probability is $\frac{1}{2}$ that he or she will pass on the freckles gene and $\frac{1}{2}$ that he or she will pass on the no-freckles gene.

Think About This Situation

Consider the chance of inheriting freckles.

a In what sense does the gene for freckles dominate the gene for no freckles?

b What is the probability that a child will have freckles if both parents do not have freckles?

c What is the probability that a child will not have freckles if each parent has one freckles gene and one no-freckles gene?

d What is the probability that a child of freckled parents also will have freckles?

INVESTIGATION 1 Multiplying Probabilities

You have found that graphical representations of data can reveal important underlying patterns and that making a "picture" of a mathematical situation can help you understand that situation better. In this investigation, you will use an *area model* to explore patterns in chance situations.

1. About half of all U.S. residents are female. According to a survey published in *USA Today*, three out of five adults sing in the shower.

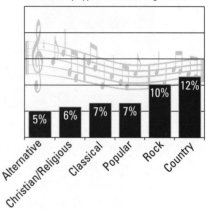

Singing in the shower

Three out of five adults say they sing in the shower. Top types of showering music:

Alternative 5%
Christian/Religious 6%
Classical 7%
Popular 7%
Rock 10%
Country 12%

Source: Guideline Research and Consulting Corporation for Westin Hotels and Resorts

a. Suppose a person from the United States is selected at random. From the information above, what do you think would be the probability that the person is a female who sings in the shower?

b. Now examine the situation using the *area model* shown to the right. Explain why there are two rows labeled "No" for "Sings in Shower" and three labeled "Yes." What assumption does this model make about singing habits of males and females?

c. On a copy of this area model, shade in the squares that represent the event of a female who sings in the shower.

d. What is the probability that a person selected at random is a female who sings in the shower?

e. What is the probability that a person selected at random is a male who does not sing in the shower?

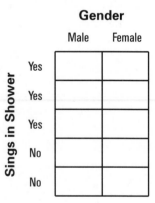

Gender

Male | Female

Sings in Shower:
Yes
Yes
Yes
No
No

2. In Lesson 1, you estimated answers to questions like the following one, involving rolling a pair of dice:

What is the probability that it takes exactly two tries to roll doubles?

Now use an area model to analyze this problem.

a. Explain why it makes sense to label the rows of the area model as shown below. On a copy of this area model, label the six columns to represent the possible outcomes on the second roll of a pair of dice.

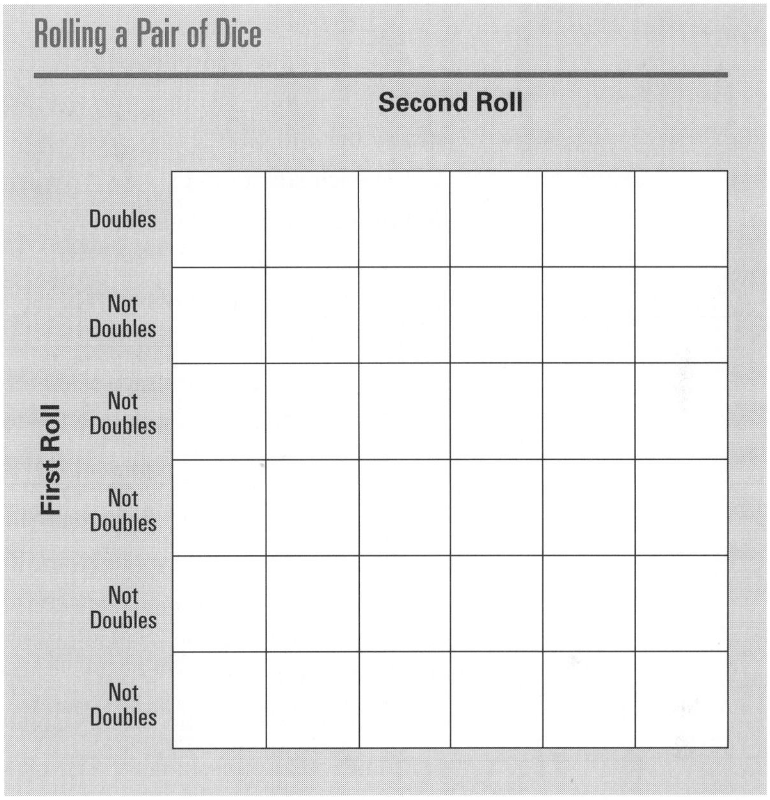

Rolling a Pair of Dice

Second Roll

First Roll

Doubles						
Not Doubles						
Not Doubles						
Not Doubles						
Not Doubles						
Not Doubles						

b. Find the squares that represent the event of not getting doubles on the first roll and getting doubles on the second roll. Shade those squares on your copy.

c. Use your area model to find the probability of not getting doubles on the first roll and then getting doubles on the second roll.

d. Use your area model to find the probability you will get doubles both times.

e. Use your area model to find the probability you won't get doubles either time.

3. Make an area model to help you determine the probabilities that a child will or will not have freckles, when each parent has one freckles gene and one no-freckles gene.

a. What is the probability that the child will not have freckles?

b. Compare your answer to Part a with your class's answer to Part c of the "Think About This Situation" on page 471.

4. Make and use area models to answer these questions:

   a. About 25% of Americans put catsup directly on their fries, rather than on the plate. What is the best estimate for the probability that *both* your school principal and your favorite celebrity put catsup directly on their fries?

   b. About 84% of Americans pour shampoo into their hand rather than directly onto their hair. What is the best estimate of the probability that *both* your teacher and the President of the United States pour shampoo into their hands before putting it on their hair?

5. Look back at the situations described in Activities 1 through 4. The pairs of events in each of those activities were (or were assumed to be) **independent events**. Knowing whether one of the events occurs does not change the probability that the other event occurs.

   a. For each situation, explain why the events are independent.

   b. For each activity, describe how to compute the probabilities without making an area model.

   c. Describe in words how to find the probability that two independent events both occur.

   d. Suppose *A* and *B* are independent events. Express your method in Part c using symbols by completing the following equation:

$$P(A \text{ and } B) = \underline{\hspace{3cm}}$$

   The notation *P*(*A* and *B*) is read "the probability of event *A* and event *B*."

Often a probability problem is easier to understand if it is written in words that are more specific than the words the original problem uses. For example, you could express and find the probability of taking exactly two tries to roll doubles in the following manner:

P(don't roll doubles on the first try and do roll doubles on the second try)

   $= P(\text{don't roll doubles on the first try}) \cdot P(\text{do roll doubles on the second try})$

   $= \left(\frac{5}{6}\right)\left(\frac{1}{6}\right)$

   $= \left(\frac{5}{36}\right)$

This example uses the *Multiplication Rule* you discovered in Activity 5 to calculate the probability that two independent events both occur.

6. Suppose Shiomo is playing a game in which he needs to roll a pair of dice and get doubles and then immediately roll the dice again and get a sum of six. He wants to know the probability that this will happen.

   a. Which of the following best describes the probability Shiomo wants to find?

   ▪ *P*(gets doubles on the first roll or gets a sum of six on the second roll)

   ▪ *P*(gets doubles on the first roll and gets a sum of six on the second roll)

   ▪ *P*(gets doubles and a sum of six)

   b. Explain why the Multiplication Rule can be used to find the probability that this sequence of two events will happen. What is the probability?

7. A modification of the game in Activity 6 involves rolling a pair of dice three times. In this modified game, Shiomo needs to roll doubles, then a sum of six, and then a sum of eleven.

a. Find the probability that this sequence of three events will happen.

b. Suppose *A*, *B*, and *C* are three independent events. Write a rule for calculating *P*(*A* and *B* and *C*) using the probabilities of each individual event.

c. Write the Multiplication Rule for calculating the probability that each of four independent events occurs.

8. For each of the following questions, explain why the events are independent. Then, use the Multiplication Rule to answer the question.

a. What is the probability that a sequence of seven flips of a fair coin turns out to be exactly HTHTTHH?

b. What is the probability that a sequence of seven flips of a fair coin turns out to be exactly TTTTTTH?

c. In the United States, about 105 boys are born for every 100 girls. What is the probability of a family having 10 boys in a row?

d. What is the probability that a family with two children will have an older girl and a younger boy? Is this probability different than the probability that the family will have an older boy and a younger girl? Explain your reasoning.

9. As a class, decide exactly how long a person's hair must be for it to be considered "long." Count the number of students in your classroom who have long hair. Count the number of girls. Count the number of students who have long hair and are girls. Suppose you select a student at random from your class.

a. What is the probability that the student has long hair?

b. What is the probability that the student is a girl?

c. Does the Multiplication Rule correctly compute the probability that the student has long hair *and* is a girl?

d. How is this situation different from previous situations in which the Multiplication Rule gave the correct probability?

10. Sometimes you want to know the probability of an event occurring when you know that another event occurs.

a. Using the data from Activity 9, find *P*(student has long hair if the student is a girl). How does this compare to *P*(student has long hair)? Are the events *having long hair* and *being a girl* independent? Why or why not?

The phrase "the probability event *A* occurs if event *B* occurs" is written symbolically as $P(A|B)$. It is sometimes read "the probability of *A* given *B*."

b. Refer to your area model for the probabilities of whether a child will have freckles if both parents have one freckles gene and one no-freckles gene. (See Activity 3 on page 473.) Compare the following:

   P(freckles gene passed from father)

   P(freckles gene passed from father|freckles gene passed from mother)

   Are the events *freckles gene passed from father* and *freckles gene passed from mother* independent?

c. Which is greater if you roll a pair of dice once: *P*(doubles) or *P*(doubles|sum is 2)? Are the events *getting doubles* and *getting a sum of two* independent?

d. If events *A* and *B* are independent, how are $P(A)$ and $P(A|B)$ related?

e. If events *A* and *B* are dependent, what, if anything, can you conclude about $P(A)$ and $P(A|B)$?

Checkpoint

In this investigation, you discovered the Multiplication Rule for independent events.

a Why does it make sense to multiply the individual probabilities when you want to find the probability that two independent events both happen?

b What is the difference between $P(A)$ and $P(A|B)$?

Be prepared to share your thinking with the entire class.

On Your Own

Think about the significance of the independence or dependence of events as you complete the following tasks.

a. While playing Monopoly, Jenny is sent to jail. She wants to know the probability that she will fail to roll doubles in three tries.

   ■ Rewrite this probability situation describing the sequence of events that are likely to occur.

   ■ Find the probability that Jenny fails to roll doubles in three tries.

   ■ Explain why you can use the Multiplication Rule for this situation.

b. Suppose you pick a high school student at random. For each of the pairs of events below, write the mathematical equality or inequality that applies:

$$P(A) = P(A|B), \ P(A) > P(A|B), \ \text{or} \ P(A) < P(A|B).$$

- *A* is the event that the student is male and *B* is the event that the student is over six feet tall.

- *A* is the event that the student is female and *B* is the event that the student has brown eyes.

- *A* is the event that the student is a member of the French club and *B* is the event that the student is taking a French class.

c. Which of the pairs of events in Part b is it safe to assume are independent? Explain your reasoning.

MORE
Modeling • Organizing • Reflecting • Extending

Modeling

1. If you completed Extending Task 1 on page 469, you are already familiar with the following famous Swedish trial: A parking officer had noted the position of the valve stems on the tires on one side of a car; upon returning later, the officer noted that the valve stems were still in the same position. The officer noted the position of the valve stems to the nearest "hour." For example, in the following picture, the valve stems are at 3:00 and at 10:00. The officer issued a ticket for overtime parking. However, the owner of the car claimed he had moved the car and returned to the same parking place.

a. Use the Multiplication Rule to estimate the probability that if a vehicle is moved, the valve stems return to the same position they had before the car was moved.

b. What assumption are you making? How can you find out if it is reasonable?

c. Do you think the man should have been issued a ticket? If you previously completed Extending Task 1 on page 469, compare your conclusions.

2. Suppose you are playing Monopoly and have been sent to jail. Recall that in Monopoly you can get out of jail by rolling doubles. If you don't roll doubles in three tries, you must pay $50 to get out of jail or use a "Get Out of Jail Free" card.

  a. What is the probability you will roll doubles and get out of jail on your first try?

  b. What is the probability you will not roll doubles on your first try and will roll doubles on your second try?

  c. What is the probability you will not roll doubles on your first try, will not roll doubles on your second try, and will roll doubles on your third try?

  d. What is the probability that you will get out of jail without having to pay $50 or use a card?

  e. What is the probability that you will have to pay $50 or use a card to get out of jail?

3. About 12.6% of the 102 million TV households in the United States watched Game 4 of the 2001 National Basketball Association championship between the L.A. Lakers and the Philadelphia 76ers. About 11.2% watched the fifth game.

  a. Why isn't it reasonable to estimate that (0.126)(0.112) or approximately 1.4% of the U.S. households watched both Game 4 and Game 5?

  b. What would be a better estimate?

4. Genetics is the study of how characteristics such as freckles or tongue roll are passed from one generation to the next. The laws of inheritance were first understood by Gregor Mendel just over 100 years ago.

  a. How many people in your group can roll their tongues up from the sides into a U-shape?

  b. Each person has two genes, one from each parent, that determine whether he or she can roll his or her tongue. The gene for tongue-rolling dominates the gene for no tongue roll. What is the probability that a child will be able to roll his or her tongue into a U-shape, if each parent has exactly one tongue-rolling gene? Explain your solution method.

  c. Suppose one parent has two tongue-rolling genes and the other parent has exactly one tongue-rolling gene. What is the probability that their child will be able to roll his or her tongue? Explain your answer.

Organizing

1. For each experiment below, find $P(A)$ and $P(A|B)$. Which of these pairs of events A and B are independent?

   a. The experiment is rolling a pair of dice once. Event A is getting doubles and event B is getting a sum of 7.

   b. The experiment is flipping a coin twice. Event A is getting a head on the second flip and event B is getting a head on the first flip.

   c. The experiment is picking a day in the year at random. Event A is getting a Sunday and event B is getting a school day.

2. If you select two random numbers that are both between 0 and 1, what is the probability that they are both greater than 0.5? You can think geometrically about this kind of problem, as shown at the right.

   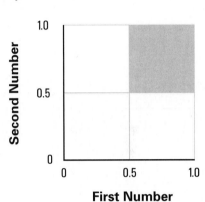

   a. Explain how the shaded region represents the event that both numbers are greater than 0.5.

   b. What is the probability that both numbers are greater than 0.5?

   c. What is the probability that at least one of the numbers is greater than 0.5?

3. Make and analyze an area model to answer the following questions.

   a. If you select two random numbers that are both between 0 and 1, what is the probability that they are both less than 0.2?

   b. If you select two random numbers that are both between 0 and 1, what is the probability that their sum is less than 1? What are the equations of the lines that border the region that represents this event?

   c. If you select two random numbers that are both between 0 and 1, what is the probability that their sum is less than 0.3? What are the equations of the lines that border the region?

   d. Al and Bill will both call Briana at a random time during their lunch hour, from 12:00 until 1:00. Each will talk to Briana for 10 minutes. What is the probability that one of them is talking to Briana when the other calls?

   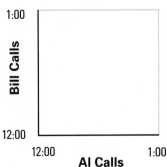

4. Consider the experiment of rolling two dice. Give examples of events *A* and *B* in which the following are true:

  a. $P(A) > P(A|B)$

  b. $P(A) = P(A|B)$

  c. $P(A) < P(A|B)$

5. Look over the probabilities you calculated in Lessons 1 and 2. If the probability that an event *A* will occur is *p* (that is, $P(A) = p$), what is the probability that the event will not occur, $P(\text{not } A)$? Explain why your conclusion makes sense.

Reflecting

1. April and May are playing Monopoly and are both in jail. April has tried twice to roll doubles and failed both times. May has tried only once, and she also was unsuccessful. Who has the better chance of rolling doubles on her next turn? Explain your reasoning.

2. In which of the following examples do you think it is reasonable to assume the events are independent?

  a. The experiment is rolling a pair of dice twice in a row. The first event is not getting doubles on the first roll. The second event is getting doubles on the second roll.

  b. The experiment is selecting two people at random. The first event is the first person pouring shampoo directly onto his or her hair. The second event is the second person pouring shampoo directly onto his or her hair.

  c. The experiment is selecting two people at random. The first event is the first person putting catsup directly on his or her fries. The second event is the second person putting catsup directly on his or her fries.

  d. The experiment is selecting one person at random. The first event is getting a person with voice (singing) training. The second event is getting a person who can play a musical instrument.

  e. The experiment is waiting for the results of next year's sports championships. The first event is the Celtics winning the NBA championship. The second event is the Reds winning the World Series.

f. The experiment is selecting a person at random. The first event is getting a person who puts catsup directly on his or her fries. The second event is getting a person who puts shampoo directly on his or her hair.

g. The experiment is selecting a pair of best friends at random from a high school. The first event is the first friend attending the last football game. The second event is the second friend attending the last football game.

3. The idea of independent events can be somewhat difficult to understand. Suppose that someone in your class has asked you to explain it. Write an explanation of the difference between independent events and dependent events. Include examples that would interest students in your high school.

4. Jesse read a survey that said that 90% of American husbands would marry the same woman again, and 72% of American wives would marry the same man again. He computed the probability that a married couple would marry each other again as follows:

P(husband would marry same wife and wife would marry same husband)

    = *P*(husband would marry same wife) · *P*(wife would marry same husband)

    = (0.90)(0.72)

    = 0.648

Is Jesse correct? Explain your reasoning.

5. Sometimes the Multiplication Rule is called the "And Rule" by students. What do you see as possible advantages and disadvantages of this alternate name?

Extending

1. In Investigation 2 of Lesson 1, three games were described for Activity 2. The rules are reproduced below. To play the games, you will need a bag with one red and one yellow marker in it and some extra red and yellow markers.

    In each game, the goal is to draw a red marker. When a red marker is drawn, the game stops and the player's score is the number of draws required. Each player starts a turn with one red and one yellow marker in the bag. The winner of the game is the person with the smallest score.

Game A: Draw until you get a red marker. Replace the marker after each draw. In addition, each time you draw a yellow marker, you must add another yellow marker to the bag before drawing again.

Game B: Draw until you get a red marker. Replace the marker after each draw. In addition, each time you draw a yellow marker, you must add a red marker to the bag before drawing again.

Game C: Draw until you get a red marker. Replace the marker after each draw, but don't add any other markers to the bag. (There are always the original two markers in the bag.)

a. For Game C, complete a probability distribution table like the one below.

Game C

Number of Draws to Get First Red	Probability
1	
2	
3	
4	
5 or more	

b. Complete a similar table for Game A.

c. Finally, complete a similar table for Game B.

d. What is the probability that you will draw a red in two draws or fewer if you are playing Game A? Game B? Game C?

e. Recall that we are considering an event in the upper 5% of a waiting-time distribution to be a rare event. Has a rare event happened if it takes 5 or more draws for Game A? For Game B? For Game C?

f. Write a report giving a complete analysis of the three games.

2. If events A and B are not independent, then the following are true.

$$P(A \text{ and } B) = P(A) \cdot P(B|A)$$
and
$$P(A \text{ and } B) = P(B) \cdot P(A|B)$$

a. Show that this rule is true for each of the following cases.

■ The experiment is rolling a pair of dice once. Event A is rolling doubles. Event B is getting a sum of 8.

■ The experiment is rolling a pair of dice once. Event A is rolling doubles. Event B is getting a sum of 7.

b. Is this rule true even if A and B are independent? Explain.

3. *Tree graphs* are a way of organizing all possible sequences of outcomes. For example, the tree graph below shows all possible families of exactly three children (with no twins or triplets). Each "G" means a girl was born, and each "B" means a boy was born. In the United States, the probability that a girl is born is approximately 49%.

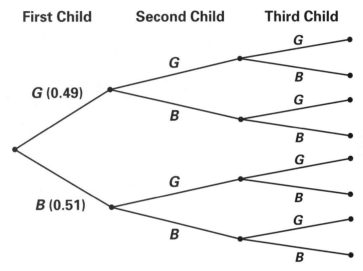

a. Use the graph to find the probability that a family of three children will consist of two girls and a boy (not necessarily born in that order).

b. Make a tree graph that shows all possible outcomes if you roll a die twice and each time read the number on top. What is the probability you will get the same number twice?

c. Make a tree graph that shows all possible outcomes if you flip a coin four times. What is the probability you will get exactly two heads?

d. Make a tree graph that shows all possible outcomes if you buy three boxes of Kellogg's® Cocoa Krispies® cereal, each containing a sticker showing one of the following: a bird of paradise, a tiger, an African elephant, and a crocodile. What is the probability you will get three different stickers?

4. In the Game of LIFE®, there are several PAY DAY spaces throughout the board. On each turn in this game, the player spins a spinner similar to the one below and moves the indicated number of spaces around the board.

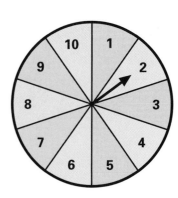

Suppose you are 5 spaces away from the next PAY DAY on the game board. You could land on the space by spinning a 5 on your next turn. Another way to land on the space is by spinning a 1 on your next turn, a 3 on the following turn, and a 1 on the turn after that.

a. Make a tree graph that shows all possible sequences of spins that would get you to this PAY DAY.

b. What is the probability that you will land on this PAY DAY space on your trip around the board?

Probability Distributions

While a prisoner of war during World War II, J. Kerrich conducted an experiment in which he flipped a coin 10,000 times and kept a record of the outcomes. A portion of the results is given in the table below.

Number of Tosses	Number of Heads
10	4
50	25
100	44
500	255
1,000	502
5,000	2,533
10,000	5,067

Source: J. Kerrich. *An Experimental Introduction to the Theory of Probability.* Copenhagen: J. Jorgenson and Co., 1964.

Think About This Situation

Think about Kerrich's results as you answer these questions.

a How many heads would you expect if you tossed a coin 100 times? 32 times? 15 times?

b After how many tosses is the number of heads in Kerrich's table closest to the expected number of heads? Furthest?

c Was the percentage of heads closer to the expected percentage of 50% after tossing 10 times or 10,000 times?

d If six Monopoly players are sent to jail, how many would you expect to get out of jail by rolling doubles on their first try? On their second try?

INVESTIGATION 1 Theoretical Waiting-Time Distributions

In Lesson 1, you explored waiting-time distributions by conducting experiments and simulations. As you probably noticed, two different groups could get quite different histograms when they constructed them from simulations. The two groups then would have different estimates of a probability. In this investigation, you will construct waiting-time distributions theoretically, so everyone should get the same answers to the probability questions related to these distributions.

If you flip a fair coin 10 times, you *expect* to get 5 heads. However, you don't get 5 heads each time you flip a fair coin 10 times. Sometimes you get fewer, as Kerrich did; sometimes you get more. In the long run, however, the average number of heads will be 5.

Expectation or *expected value* is another word for the *theoretical average*. For example, if you flip a coin 5 times, you might say in ordinary language that you expect to get 2 or 3 heads. However, in mathematics you should say that you expect to get 2.5 heads, because *on the average* that is the number of heads you will get.

1. Imagine 36 students are playing modified Monopoly in a class tournament. All are sent to jail. A student must roll doubles to get out of jail. (There is no other way out.)

   a. How many of the 36 students do you expect to get out of jail by rolling doubles on the first try? (Remember that the word "expect" has a mathematical meaning.) How many students do you expect to remain in jail?

   b. How many of the remaining students do you expect to get out of jail on the second try? How many students do you expect to remain in jail then?

   c. How many of the remaining students do you expect to get out of jail on the third try? How many students do you expect to remain in jail then?

   d. Complete a table like the one at the top of the next page. Round numbers to the nearest hundredth. The first three lines should agree with your answers to Parts a–c.

Rolling Dice to Get Doubles

Number of Rolls to Get Doubles	Expected Number of Students Released on the Given Number of Rolls	Expected Number of Students Still in Jail
1		
2		
3		
4		
5		
6		
7		
8		
9		
10		
11		
12		

   e. What patterns of change do you see in this table? If possible, describe each pattern using the idea of *NOW* and *NEXT*.

The table you created shows a **theoretical distribution**. The table never really can be completed, however, as the rows should be continued indefinitely.

2. Use the theoretical distribution to reason beyond the table entries.

   a. How many of the 36 students do you expect to be in jail after 12 tries to roll doubles?

   b. Add a "13 or more" row to your table and write the expected number of students in the appropriate place.

   c. If you had started with 10,000 people instead of 36, how many of them would you expect to need 13 or more rolls to get out of jail?

3. Make a histogram of the "expected number of students released on the given number of rolls" from the frequency table you constructed in Activity 1.

   a. Compare this histogram to the one you constructed following your class's simulation of this situation in Activity 1 of the first investigation from Lesson 1 (page 457–458).

   b. Examine your histogram of the theoretical distribution. The height of each bar is what proportion of the height of the bar to its left?

   c. Using your histogram, estimate the average of the distribution.

   d. Calculate the average number of rolls of the dice it takes to get doubles. Compare your calculated average to your estimate in Part c.

4. If there were 1,000 people (rather than 36) who had been sent to jail in Activity 1, how would the histogram change? How would the average change?

5. Thirty percent of "M&M's"® Plain Chocolate Candies are brown. Suppose each of 1,000 students removes candies one at a time from a large bag until he or she gets a brown one.

   a. How many students do you expect to get a brown candy on the first try?

   b. Make a table like the one in Activity 1. Give the table 10 rows, and complete each row.

   c. How many of the 1,000 students do you expect to need 11 or more draws to get a brown candy?

   d. Make a histogram of the distribution shown in your table.

   e. Estimate the average number of draws to get a brown candy from your histogram. Then calculate the average using an appropriate formula.

Checkpoint

Summarize key ideas about theoretical waiting-time distributions.

a How is the word "expect" used differently in mathematics than in everyday life?

b Suppose the probability of an event is p. How many times would you expect this event to happen in a series of n independent trials?

c Suppose the probability of an event is p and you have made a histogram of the waiting-time distribution for the event. How is the height of each bar of the histogram related to the height of the bar to its left?

Be prepared to share your ideas with the whole class.

▶On Your Own

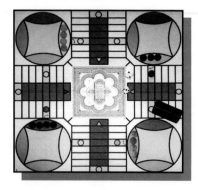

In the game of Parcheesi®, based on the Indian game pachisi, a player cannot move a pawn for the first time until he or she rolls a five with a pair of dice. The five may be on either die, or the five may be the sum of both dice.

   a. What is the probability a player can move a pawn on the first roll of the dice?

   b. If four players begin a game of Parcheesi, how many do you expect to move a pawn on their first roll of the dice?

   c. How many of the four players do you expect to require two rolls of the dice in order to move a pawn?

MORE

Modeling

1. If there is a 40% chance of rain today, it means that it rained on 40% of the days in the past that had weather conditions similar to those today.

   a. On 14 different days, the weather report says there is a 40% chance of rain. On how many of these days do you expect it to rain? On how many of these days do you expect it not to rain?

   b. On 20 different days, the weather report says there is a 50% chance of rain. It actually rained on 9 of those days. Do you think the meteorologist did a good job of predicting rain? Explain.

2. Tay-Sachs disease results when a baby inherits the Tay-Sachs gene from both parents. (A person who can pass the gene to the next generation is called a carrier.) Since the disease results in early death, no adults carry two Tay-Sachs genes. A college biology textbook says that "about 1 in 30 American Jews is a carrier, which would result in about 1 Tay-Sachs child in 3600 Jewish births." (Source: Audesirk, G. and T. *Biology: Life on Earth*. New York: Macmillan, 1989.) Explain how the 1 in 3600 figure was computed.

3. In each Kellogg's® Cocoa Krispies® cereal box, there was a sticker of either a bird of paradise, a tiger, an African elephant, or a crocodile. Assume that the stickers were placed randomly into the boxes. Thirty-six people are buying boxes, each trying to get a tiger.

   a. Make a theoretical waiting-time distribution table and histogram of the number of boxes purchased to get a tiger sticker. Place the numbers 1 to 11 in your table followed by a "12 or more" row.

   b. How many of these people do you expect would need to purchase more than six boxes of the cereal?

   c. On the average, how many boxes of cereal need to be purchased to get a tiger sticker?

4. Imagine 1,024 students in a school auditorium all standing up. Each student flips a coin. The students who get heads on this flip sit down. Each student who remains standing flips a coin a second time. The students who get heads on this flip sit down. The coin flips continue until all students are seated.

   a. Make a frequency table that shows what you expect to happen in this experiment.

   b. Make a histogram from your frequency table.

   c. What is the average number of flips required until a student sits down?

   d. What is the probability that it will take two flips or fewer to get a head? What is the probability that it will take more than two flips?

   e. How many times would a student have to toss the coin without getting a head before you would say a rare event has occurred?

   f. How many rare events do you expect to occur among the 1,024 students?

5. The player with the highest field goal percentage in the history of the National Basketball Association (NBA) is Artis Gilmore. In his career in the NBA, Gilmore attempted 9,570 field goals and made 5,732 of them.

   a. What was Gilmore's field goal percentage?

   b. During a typical game, Gilmore might attempt 25 field goals. In a typical game, how many field goals would you expect Gilmore to make?

The NBA player with the highest lifetime free throw percentage is Mark Price. Price had a free throw "percentage" of 0.904. He made a total of 2,135 free throws.

   c. Why do you think the word *percentage* is in quotation marks above?

   d. How many free throws did Price attempt?

   e. How many free throws would you expect Price to make in 50 attempts?

   f. Write an equation that relates the number of free throws T expected for a player who makes A attempts and whose free throw percentage is p.

Organizing

1. Refer to the waiting-time distribution for a brown candy in Activity 5 of Investigation 1 (page 488).

   a. Make a box plot of these waiting times.

   b. What is the median number of candies drawn to get the first brown one?

   c. How does the shape of the box plot reflect the shape of the histogram? In which direction is the distribution skewed?

2. When would the mean be a good measure of the center of a waiting-time distribution? When would the median be a good measure?

3. Explain why the histogram of a waiting-time distribution does not have line symmetry.

4. Refer to Activity 1 (page 486) of Investigation 1 for this task.

   a. Make a scatterplot of the (*number of rolls to get doubles*, *expected number of students who get out of jail on this roll*) data.

   b. Find an algebraic model of the form "$y = \ldots$" that is a good fit for these data.

   c. Let *NOW* be the number of people who are expected to get doubles on a roll, and let *NEXT* be the number of people who are expected to get doubles on the next roll. Write an equation relating *NOW* and *NEXT*.

5. Make a box plot for your distribution from Modeling Task 3 (getting a tiger sticker). Using the same number line, make box plots for the distribution from Modeling Task 4 and for the distribution from Activities 1 and 2 (rolling doubles) of Investigation 1. Write a few sentences about what you can learn about waiting-time distributions from examining these box plots.

Reflecting

1. It is quite common for a person's first guess about a probabilistic situation to be wrong. Part of learning probability is learning to be wary of your first reaction and learning how to check whether your first reaction is correct. What have you found most surprising so far in this unit?

2. List some businesses or jobs in which the manager should be interested in waiting-time distributions.

3. Marina and Jamie are playing a game in which the first person to roll doubles wins. Marina has had 10 turns and hasn't rolled doubles yet. Marina says, "I'm due to get doubles on my next roll."

   a. Explain what Marina means by this statement.

   b. Is Marina correct? Why or why not?

   c. Design an experiment to show Marina that she is no more likely to roll doubles on her 11th roll than she was on her 1st roll.

4. Board games involving chance have a long history of providing recreation for people from many different cultures.

   a. Investigate the history of backgammon, Parcheesi®, Senet (an Egyptian game), or a similar game that you have played with your family or friends.

   b. Review the rules of the game.

   c. Write several questions about the probabilities involved in the game.

   d. Answer one of your questions either theoretically or by using a simulation to estimate the probability.

Extending

1. What are some characteristics that theoretical waiting-time distributions for independent trials have in common?

2. In this task you will find the average of another type of frequency distribution. This distribution is called a **binomial distribution**. "Binomial" means "having two names."

   On each flip of a coin, you can describe the outcome using one of two names: "heads" or "tails." In a waiting-time distribution, you might count the *number of trials* until you get the first head. In a binomial distribution, you count the *number of heads* in a fixed number of independent trials.

   a. Flip 11 coins and count the number of heads. Add the result to a copy of the frequency table below.

Number of Heads	Frequency
0	1
1	0
2	4
3	5
4	16
5	18
6	27
7	15
8	5
9	3
10	1
11	0
Total	

b. Repeat this experiment four more times until you have a total of 100 frequencies. Does it make a difference if you flip one coin 11 times and count the number of heads or flip 11 coins at one time and count the number of heads?

c. Make a histogram from the frequency distribution.

d. How does the shape of the histogram differ from that of the waiting-time distributions? Does this histogram have line symmetry?

e. What is the average number of heads in the 100 experiments? Theoretically, how many heads would you expect in 11 flips of a coin?

f. Use your frequency table to estimate the probability that if a couple has 11 children, they are either all girls or all boys. What assumptions are you making?

3. According to the National Center for Statistics and Analysis, about 30% of all traffic fatalities are due to too fast or unsafe speeds. Suppose that in a certain county, 40 of the last 50 traffic fatalities were due to too fast or unsafe speeds. In this task, you will investigate whether it is reasonable to attribute this result to chance variation or whether this county should look for some other explanation.

a. Assume that the probability is 0.30 that a traffic fatality is due to too fast or unsafe speeds. Design a simulation using random digits to determine the number of traffic fatalities out of 50 that are due to too fast or unsafe speeds.

b. Repeat your simulation enough times so that you feel confident in deciding whether 40 out of 50 traffic fatalities is a result that reasonably could be attributed to chance or whether the county should look for another explanation. What is your conclusion? How many times did you repeat the simulation?

c. Would your conclusion be different if the county had had 30 such traffic fatalities out of 50?

4. The formula below gives the probability of getting exactly x heads if a coin is flipped n times.

$$P(x) = \frac{n!}{x!(n-x)!}\left(\frac{1}{2}\right)^n$$

The symbol $n!$, read "n factorial," is defined as follows:

$0! = 1$

$1! = 1$

$2! = 2 \cdot 1 = 2$

$3! = 3 \cdot 2 \cdot 1 = 6$

$4! = 4 \cdot 3 \cdot 2 \cdot 1 = 24$

and so on.

a. If you flip a coin 11 times, what is the probability of getting exactly 4 heads?

b. Use the formula to complete the table below for the experiment of flipping a coin 11 times.

Number of Heads	Probability
0	
1	
2	
3	
4	
5	
6	
7	
8	
9	
10	
11	

c. Make a graph of the distribution in this table. How does the graph compare to your histogram from Extending Task 2?

d. How does the formula incorporate the Multiplication Rule?

INVESTIGATION 2 Probability Distributions and Rare Events

The table and histogram below give the proportion of families in the United States that are a given size.

Size of Family	Proportion
2	0.43
3	0.23
4	0.20
5	0.09
6	0.03
7 or more	0.02
Total	1.00

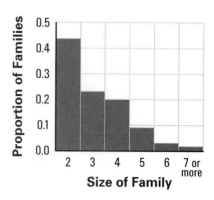

Source: U.S. Bureau of the Census, *Statistical Abstract of the United States: 2000* (120th edition). Washington, DC, 2000.

1. As a class, discuss the following questions.

   a. If you were to pick a family at random from the United States, what is the probability that it would have four people in it? What is the probability it would have four or fewer people in it?

   b. The family pictured below is remarkable in at least two ways. Do you think a family of this size is a rare event? Explain your reasoning.

   c. In addition to the size, what else seems remarkable about this family? Do you think this is a rare event? What do you think is the probability of this event occurring, given that there are nine children in the family?

In the remainder of this investigation, you will learn to construct and use **probability distributions**. A probability distribution tells you at a glance the probabilities associated with all possible events. For example, the table of family sizes on page 495 tells you the probability that a family selected at random in the United States will have exactly five members. It was constructed after a census that tried to count all families in the United States.

Probability distribution tables also can be constructed for theoretical events. Shown below is the probability distribution table and graph for the experiment of rolling a die and reading the number on the top. Note that the shape of the graph of this probability distribution is *rectangular*.

Number on Die	Probability
1	$\frac{1}{6}$
2	$\frac{1}{6}$
3	$\frac{1}{6}$
4	$\frac{1}{6}$
5	$\frac{1}{6}$
6	$\frac{1}{6}$
Total	$\frac{6}{6}$

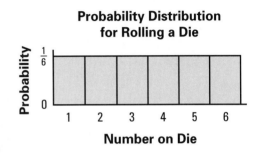

Probability Distribution for Rolling a Die

2. Imagine the experiment of rolling a tetrahedral die (with four faces).

   a. Make a probability distribution table and graph for this situation.

   b. What is the shape of your graph?

3. Consider again the experiment of rolling two standard dice and adding the two numbers on the tops of the dice. (See Lesson 1, page 458.)

   a. Complete a probability distribution table like the one at the right. Write the probabilities as fractions.

Sum of Two Dice	Probability
2	
3	
4	
5	
6	
7	
8	
9	
10	
11	
12	
Total	$\frac{36}{36}$

b. Below is a partially-completed graph of the probabilities in your table. Copy and complete the graph.

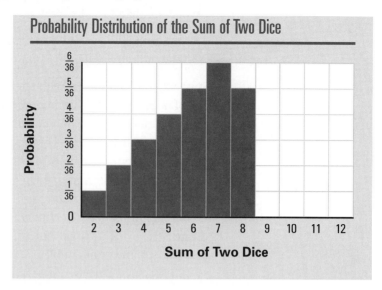

Probability Distribution of the Sum of Two Dice

c. The shape of the distribution in Part b is *triangular*. What kinds of situations or experiments can be expected to give probability distributions that are rectangular? Triangular?

Writing with words about probability distributions such as the one above can get somewhat complicated, so a more compact mathematical notation often is used. For example, the mathematical notation

$$P(\text{sum} \geq 7) = \frac{21}{36}$$

says exactly the same thing as

the probability that the sum of the two dice is seven or greater equals $\frac{21}{36}$.

If it is understood that you are talking about the sum of two dice, you can use the notation

$$P(5) = \frac{4}{36}$$

to mean

the probability that the sum of the two dice is 5 equals $\frac{4}{36}$.

4. This activity will help you become comfortable using this new notation. Refer to the probability distribution in Activity 3.

a. Find each of the following probabilities about the sum of the two dice.

- $P(4)$
- $P(\text{sum} > 5)$
- $P(\text{sum} \geq 5)$
- $P(7 \text{ or } 11)$

b. Rewrite each sentence below using mathematical notation. Then decide if the statement is true or false.

- The probability that the sum is more than 12 equals 0.
- The probability that the sum is 2 equals $\frac{1}{36}$.
- The probability that the sum is an even number equals $\frac{18}{36}$.

5. Now that you are familiar with the idea of a probability distribution, you can construct the theoretical waiting-time distribution for the number of rolls of a pair of dice until doubles appear. Again, imagine that all students in your class are playing modified Monopoly. All are sent to jail. To get out of jail a student must roll doubles.

a. Copy the following waiting-time probability distribution table. You will complete the "Probability" column as you do Parts b–h. Report all probabilities as fractions.

Number of Rolls to Get Doubles	Probability
1	
2	
3	
4	
5	
6	
7	
8	
9	
10	
11	
12	

b. What is the probability of getting doubles on the first roll? Enter your answer as a fraction on the first line of the table.

c. To get out of jail on the second roll, two events must happen. You must not roll doubles on the first roll *and* you must roll doubles on the second roll.

■ What rule of probability can you use to compute $P(A$ and $B)$ when A and B are independent events?

■ What is the probability of not getting doubles on the first roll and getting doubles on the second roll? Write your answer on the second line of the table.

d. To get out of jail on the third roll, three events must happen.

■ What are these three events?

■ What is the probability all three events will happen? Write your answer on the third row of the table.

e. To get out of jail on the fourth roll, four events must happen.

■ What are these four events?

■ What is the probability all four events will happen? Write your answer on the fourth row of the table.

f. Explain why your work for Parts b–e can be summarized in the following manner.

$$P(1) = \tfrac{1}{6}$$
$$P(2) = \left(\tfrac{5}{6}\right)\left(\tfrac{1}{6}\right)$$
$$P(3) = \left(\tfrac{5}{6}\right)^2\left(\tfrac{1}{6}\right)$$
$$P(4) = \left(\tfrac{5}{6}\right)^3\left(\tfrac{1}{6}\right)$$

g. What patterns do you see in the above equations? Compare your patterns with those of other groups.

■ What is $P(5)$? $P(6)$?

■ What is $P(x)$? That is, what is the probability the first doubles will appear on the xth roll of the dice?

h. Use your general formula in Part g to complete the rest of the probability distribution table. Don't bother multiplying the fractions for now.

i. Express the probabilities as decimals and then make a graph of the probability distribution table.

j. How does this graph of the theoretical waiting-time probability distribution compare with the histogram you produced for Activity 1 of Investigation 1 in Lesson 1 (page 458)?

6. Recall that 30% of "M&M's"® Plain Chocolate Candies are brown. Find the first five entries of the table for the waiting-time probability distribution for drawing a brown candy.

Examine each of the probability distributions shown below.

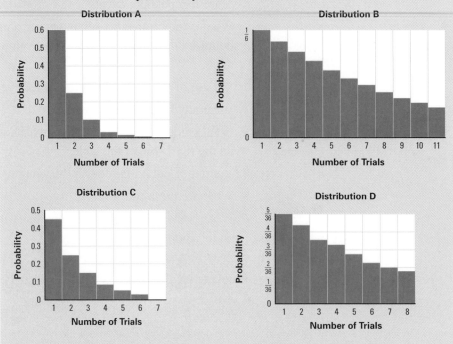

a Match the following descriptions to the probability distributions.

- The experiment is rolling a die until a 6 appears.

- The experiment is rolling two dice until a sum of 6 appears.

- The experiment is counting the days on which the weather report says there is a 60% chance of rain until there is a rainy day.

- The experiment is selecting a person at random until one with type O blood appears. (About 45% of the U.S. population has type O blood.)

b What is the height of the second bar (representing 2 trials) in each case? The third bar?

Be prepared to explain your description-graph matches to the class.

On Your Own

Use your completed probability distribution table in Activity 5, page 498, to find the following probabilities. "Rolls" stands for the number of rolls to get doubles.

a. $P(\text{rolls} \leq 3)$

b. $P(\text{rolls} > 3)$

c. $P(\text{rolls} = 7)$

---

7. Shown below is the probability distribution of the waiting time for doubles, with the probabilities expressed as decimals rounded to the nearest thousandth.

Number of Rolls	Probability	Number of Rolls	Probability
1	0.167	12	0.022
2	0.139	13	0.019
3	0.116	14	0.016
4	0.096	15	0.013
5	0.080	16	0.011
6	0.067	17	0.009
7	0.056	18	0.008
8	0.047	19	0.006
9	0.039	20	0.005
10	0.032	21 or more	0.025
11	0.027		

a. What is the probability it will take 19 or more rolls to get out of jail? Is it a rare event to take 19 rolls of the dice to get out of jail? (Recall that we are considering a rare event to be one in the upper 5% of the distribution.)

b. Suppose in playing the modified game of Monopoly, Michael is still in jail after trying 8 times to roll doubles. Has a rare event occurred?

8. Complete a probability distribution table like the one below for the experiment of flipping a coin until a head appears. Complete the "Probability" column using decimals rounded to the nearest thousandth.

Number of Flips to Get a Head x	Probability $P(x)$
1	
2	
3	
4	
5	
6	
7	
8	
9	
10	

a. Make a graph of your distribution.

b. What is the probability that the first head occurs on the second flip?

c. What is the formula for $P(x)$, the probability of getting the first head on flip number x?

d. If Scott requires 5 flips to get a head, has a rare event occurred? Explain.

e. If Michele flips a coin 8 times before the first head appears, has a rare event occurred? Explain.

Checkpoint

Refer to the table of family sizes and the family pictured at the beginning of this investigation (page 495).

ⓐ What family sizes would qualify as rare events? Explain.

ⓑ Suppose that the family pictured had planned to have children until they had a boy.

- Is a family of nine girls and no boys a rare event? Explain your reasoning.

- If the parents have another child, is it more likely to be a boy or a girl? Justify your answer.

Be prepared to share your thinking with the entire class.

On Your Own

Consider an experiment in which a blood bank is testing people at random until it finds a person with type O blood. About 45% of the U.S. population has type O blood.

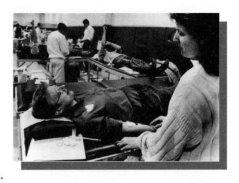

a. Make a probability distribution table for this experiment. Number the first four rows in your table and follow them with a row for 5 people or more.

b. Write a formula for $P(x)$, the probability that the xth person tested is the first with type O blood.

c. Suppose the 18th person tested were the first with type O blood. Would this be considered a rare event? Explain your reasoning.

MORE
Modeling • Organizing • Reflecting • Extending

Modeling

1. Twenty percent of "M&M's"® Plain Chocolate Candies are red.

a. Make a probability distribution table for the experiment of drawing a candy until a red one appears. Use the numbers 1 through 9 in your table and end it with a "10 draws or more" row.

b. Make a graph of your distribution.

c. Write a formula for $P(x)$, the probability of getting the first red candy on the xth draw.

d. How many candies would you have to draw before getting the first red in order for a rare event to have occurred?

2. Refer back to Modeling Task 3 on page 489.

a. Write a formula for $P(x)$, the probability of getting a tiger sticker on the xth purchase.

b. How many boxes of Kellogg's® Cocoa Krispies® cereal would a person have to buy without getting a tiger sticker before a rare event would have occurred?

3. In an episode of a television show, a man receives an anonymous letter that correctly predicts the outcome of a sports event. In the next four weeks, similar letters arrive, each making a prediction that turns out to be correct. The final letter asks the man for money before he receives another prediction. The whole thing turns out to be a scam.

   Two versions of the first letter had been sent out, each to a large number of people. Half of the people received letters that predicted Team A would win, and half of the people received letters that predicted Team B would win. Those people who received letters with the correct prediction were sent letters the second week. Again half of the letters predicted Team C and half predicted Team D.

   How many letters should have been sent out the first week so that exactly one person would be guaranteed to have all correct predictions at the end of the five weeks?

4. Now that you can describe waiting-time probability distributions with an algebraic formula, you can use a calculator to help analyze situations modeled by these distributions. A helpful calculator procedure to produce a waiting-time probability distribution table makes use of a sequence command found on some graphing calculators. The command is usually of the form seq(*formula*, *variable*, *begin*, *end*, *increment*). The following keystroke procedure uses A for the variable and 1 for both the beginning value and the increment. (The exact structure of this command and how to access it vary among calculator models. You may need to refer to the manual for your calculator.)

   First you need to access the sequence command. For example, from the home screen on a TI-83, press [2nd] [STAT] [▶] 5. Then, complete the command by entering the following example:

   (5 ÷ 6) ^ (A − 1) × 1

   ÷ 6 , A , 1 , 12 , 1)

Store the result in a data list for easier access. The following are sample display screens for this procedure:

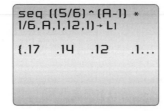
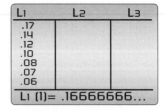

The data list now holds the first 12 table entries of the waiting-time probability distribution for rolling doubles. (Note that the decimal display was set to show only two digits.)

a. Compare the entries produced by this procedure with those you calculated in Activity 5 of Investigation 2 (page 498).

b. Compare the sequence command with the general formula you wrote in Activity 5, Part g of Investigation 2. How are they similar and how are they different?

c. Modify the sequence command to produce the first ten table entries of the waiting-time probability distribution for drawing a brown "M&M's"® Chocolate Candy. Compare the first five entries with the entries you calculated in Activity 6 of Investigation 2 (page 499).

d. Write a summary of what you need to know in order to use this procedure for producing a waiting-time probability distribution. How do you figure out the needed formula?

e. How could this calculator procedure help you analyze questions about rare events associated with waiting-time probability distributions?

Organizing

1. In this task, you will use algebraic notation to construct a formula for waiting-time distributions. In this general case, use the letter p to stand for the probability of getting the waited-for event. (In the experiment of waiting for doubles on a pair of dice, $p \approx 0.17$ on each trial. In the experiment of flipping a coin until a head appears, $p = 0.5$ on each trial.)

   a. The first row of the table below gives the probability that the waited-for event will occur on the first trial. Make a copy of this table and then fill in the first row.

Number of Trials x	Probability $P(x)$
1	
2	
3	
4	
5	
x	

   b. What is the probability of *not* getting the waited-for event on the first trial?

   c. What is the probability of not getting the waited-for event on the first trial and then getting it on the second trial? Fill in the second row of the table.

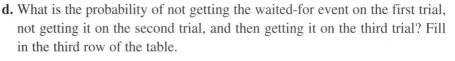
d. What is the probability of not getting the waited-for event on the first trial, not getting it on the second trial, and then getting it on the third trial? Fill in the third row of the table.

e. Finish filling in the rows of the table. Write a general formula for describing waiting-time distributions.

f. How is your general formula for Part a like the equation for an exponential model? How is it different?

2. Use your formula from Organizing Task 1 to answer these questions.

   a. The Current Population Survey of the U.S. Census Bureau recently found that 25.6% of adults over the age of 25 in the United States have four or more years of college. If a line of randomly-selected U.S. adults over the age of 25 is walking past you, what is the probability that the fifth person to pass will be the first with four or more years of college?

   b. What is the probability that parents would have seven boys in a row before having a girl? (P(boy) = 0.51) Assume births are independent.

3. Consider the median of a probability distribution.

   a. Find the median of the waiting-time-for-doubles probability distribution produced in Activity 5 of Investigation 2 (page 498). In this situation, what does the median tell you?

   b. Describe how to find the median of any probability distribution.

4. Use *NOW* and *NEXT* to describe the relationship between successive entries in the probability distribution table for waiting for doubles. (See Activity 5 of Investigation 2 on page 498.)

5. In this task, you will explore some of the geometry and algebra connected with the probability distribution for the sum of two dice.

   a. Plot the points that represent the probability distribution for the sum of two regular dice: $\left(2, \frac{1}{36}\right), \left(3, \frac{2}{36}\right), \left(4, \frac{3}{36}\right), \ldots, \left(12, \frac{1}{36}\right)$.

   - Write a single equation whose graph fits the pattern of points for $x = 2, 3, 4, 5, 6, 7$.

   - Write another single equation whose graph fits the pattern of points for $x = 7, 8, 9, 10, 11, 12$.

   - How are the slopes of these two graphs related?

   - Use absolute value to write one equation whose graph fits the pattern of all 11 points.

b. Graph the points that represent the probability distribution for the sum of two tetrahedral dice: $\left(2, \frac{1}{16}\right), \left(3, \frac{2}{16}\right), \left(4, \frac{3}{16}\right), \ldots, \left(8, \frac{1}{16}\right)$.

■ Write a single equation whose graph fits the pattern of points for $x = 2, 3, 4, 5$.

■ Write another single equation whose graph fits the pattern of points for $x = 5, 6, 7, 8$.

■ How are the slopes of these two graphs related?

■ Use absolute value to write one equation whose graph fits the pattern of all 7 points.

c. Write an equation whose graph fits the pattern of the points that represent the probability distribution for the sum of two dice, each with n sides.

Reflecting

1. The waiting-time distribution below was constructed by a computer simulation. How could you use it to estimate p, the probability of the event occurring on each trial? What is your estimate of p?

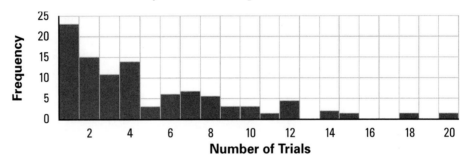

2. You have seen that if the experiment is drawing "M&M's"® Chocolate Candies until a brown one appears, the probability of getting the first brown on the 6th try is $(0.70)^5(0.30)$. Write a paragraph explaining to a friend who has been absent from school why this makes sense.

3. Explain why the heights of the bars are decreasing in a waiting-time probability distribution.

4. What are other situations that could be modeled by the probability distribution you constructed for the experiment of flipping a coin until a head appears?

5. The table below gives the mileage at which each of 191 buses had its first major motor failure.

Mileage before Failure	Number of Buses
0 to 19,999	6
20,000 to 39,999	11
40,000 to 59,999	16
60,000 to 79,999	25
80,000 to 99,999	34
100,000 to 119,999	46
120,000 to 139,999	33
140,000 to 159,999	16
160,000 and up	4
Total	191

Source: Mudholkar, G.S., D.K. Srivastava, and M. Freimer. "The Exponential Weibull Family: A Reanalysis of the Bus-Motor-Failure Data." *Technometrics* 37 (Nov. 1995): 436–445.

a. Make a histogram of this distribution.

b. Is the shape of the distribution the same as that of other waiting-time distributions you have seen? Explain why this makes sense.

Extending

1. Let *A* be the event of requiring exactly 5 rolls of a pair of dice to get doubles for the first time. Let *B* be the event that the first four rolls of a pair of dice aren't doubles. Find the following probabilities:

a. $P(A)$ **b.** $P(B)$

c. $P(A|B)$ **d.** $P(B|A)$

e. $P(A \text{ and } B)$ **f.** $P(A \text{ or } B)$

2. Make a probability distribution table for the *larger* of the two numbers when two dice are rolled. If both dice show the same number, that number is the larger one.

a. What is $P(5)$?

b. What is $P(\text{larger number} \le 2)$?

c. Make a graph of the probabilities in your probability distribution table.

d. Describe the shape of this distribution.

3. Make a probability distribution table for the absolute value of the difference of the numbers when two dice are rolled.

   a. What is $P(3)$?

   b. What is $P(1, 2, \text{ or } 3)$?

   c. Make a graph of the probabilities in your probability distribution table.

   d. Describe the shape of this distribution.

4. Shown below is the probability distribution table for two six-sided, nonstandard dice. Notice that these nonstandard dice have the same probability distribution as that of two regular dice.

Sum of Two Dice	Probability
2	$\frac{1}{36}$
3	$\frac{2}{36}$
4	$\frac{3}{36}$
5	$\frac{4}{36}$
6	$\frac{5}{36}$
7	$\frac{6}{36}$
8	$\frac{5}{36}$
9	$\frac{4}{36}$
10	$\frac{3}{36}$
11	$\frac{2}{36}$
12	$\frac{1}{36}$
Total	$\frac{36}{36}$

Using positive whole numbers, label the faces of these two nonstandard dice. The two dice may be different from one another and numbers may be repeated on the faces of a die.

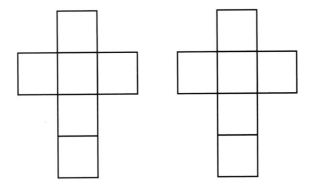

Expected Value of a Probability Distribution

Have you ever watched old gangster movies? If so, you may have heard talk of the "numbers racket." This is an illegal gambling game played in many cities. The player pays $1.00 and picks a number from 000, 001, 002, …, 999. The winning number is one that everyone can check, but no one can control, such as the last three digits of the Dow Jones Industrial Average for the day. Players who select the winning number get about $600 for a $1.00 ticket.

Wednesday's markets

Stocks pulled back from Tuesday's gains, with the Nasdaq leading the decline amid renewed profit jitters in the tech sector.

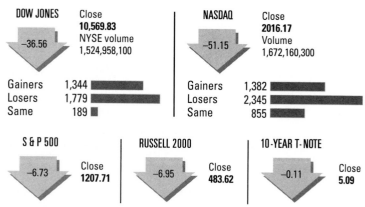

DOW JONES		NASDAQ	
−36.56	Close **10,569.83** NYSE volume 1,524,958,100	−51.15	Close **2016.17** Volume 1,672,160,300

	DOW JONES		NASDAQ
Gainers	1,344	Gainers	1,382
Losers	1,779	Losers	2,345
Same	189	Same	855

S & P 500		RUSSELL 2000		10-YEAR T-NOTE	
−6.73	Close **1207.71**	−6.95	Close **483.62**	−0.11	Close **5.09**

Source: *Chicago Tribune*, July 18, 2001.

Think About This Situation

Consider the mathematics behind the "numbers racket."

a In the long run, will a player tend to make or lose money?

b For a player to come out even in the long run, how much should a ticket cost?

c On average, how many games would a player have to play before winning a game?

INVESTIGATION 1 What's a Fair Price?

In mathematics, a **fair price** for a game is the price that should be charged so that, in the long run, the players expect to come out even. (That is, the players expect to win as much as they are charged to play.)

1. Suppose your school decides to hold a raffle. The prizes in the raffle will be a microwave that costs $400, a video game that costs $100, and a bicycle that costs $175. Exactly 2,000 tickets will be sold.

   a. What is a fair price for a ticket?

   b. Write a procedure for finding the fair price of a raffle ticket.

2. At a fund-raising carnival for a service organization, Renee is trying to get Leroy to play a game she has invented. Leroy would spin the spinner shown below and get a gift certificate worth the amount indicated. The organization charges $5 to play this game.

   a. How much would Leroy expect to win in 100 games? How much would Leroy have to pay to play 100 games? What is Leroy's expected net earnings?

   b. Explain why Leroy should or should not play this game.

   c. Leroy thinks he should play this game because he has two chances of winning, one chance of coming out even, and only one chance of losing. What would you say to him?

   d. Design a spinner that would make $5 a fair price to charge to play.

3. In roulette, a wheel and ball spin around in opposite directions. When they stop, the ball has an equal chance of landing in any one of the slots. Roulette wheels have 18 red slots, 18 black slots, and 2 green slots. Suppose a player bets on red. If the ball lands in a red slot, the player wins $2. The price to play is $1.

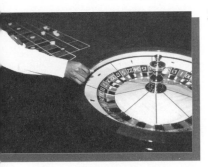

   a. Is $1 a fair price for playing roulette? Why or why not?

   b. On every 1,000 bets of $1 on red in roulette, how much money does a casino expect to make?

4. The National Center for Health Statistics reported that there were 118 deaths in 1998 for every 100,000 males aged 15–24 in the United States. There were 43 deaths per year for every 100,000 females in that age group. (Source: U.S. Bureau of Statistics, *Statistical Abstract of the United States: 2000*, 120th edition. Washington, D.C., 2000.) Use these statistics to answer the following.

a. Ignoring all factors other than gender, what would be the fair price for an insurance company to charge to insure the life of a male in that age group for one year for $50,000?

b. Ignoring all factors other than gender, what would be the fair price for an insurance company to charge to insure the life of a female in that age group for one year for $50,000?

c. If the insurance company is not allowed to have different rates for each gender, what would be the fair price for a $50,000 policy for one year? Assume that the same number of insurance policies are sold to males as to females.

d. Compare the procedure you used to get your answer to Parts a and b with your procedure in Part b of Activity 1.

e. In what ways is insurance similar mathematically to a raffle? In what ways is it different?

f. Do you think insurance companies actually charge the fair price for a policy? Explain your thinking.

Checkpoint

In this investigation, you explored how to compute the fair price for games and insurance.

ⓐ What is the relationship between the fair price of a game and the average winnings of a player in the long run?

ⓑ Describe a general procedure for finding the fair price of raffle or lottery tickets.

ⓒ Describe a general procedure for finding the fair price of an insurance policy.

ⓓ What would be a fair price for a ticket in the "numbers racket" game described at the beginning of this lesson?

Be prepared to compare your thinking and procedures with those of other groups.

On Your Own

Apply your method of finding a fair price to the two situations below.

a. The prizes in a raffle are ten $15 CDs and one $500 stereo. If 1,000 raffle tickets will be sold, what is the fair price for a ticket?

b. According to the Youth Risk Behavior Survey, about 33% of high school students reported that they had some property stolen or deliberately damaged at school within the previous year. Suppose the average value of the stolen or damaged property in that year was $30. Assuming these statistics stay the same each year, what would be the fair price to charge a student who wanted to be insured against theft or damage for one year of high school?

INVESTIGATION 2 Fair Price and Expected Value

You have learned how to construct and analyze probability distributions. In this investigation, you will learn a method of computing the fair price of a game if you are given the probability distribution of the prizes.

1. Examine the probability distribution table below.

Prize Value x	Probability $P(x)$
$1	$\frac{1}{6}$
$2	$\frac{1}{6}$
$3	$\frac{1}{6}$
$4	$\frac{1}{6}$
$5	$\frac{1}{6}$
$6	$\frac{1}{6}$
Total	$\frac{6}{6}$

a. What is the fair price for a game that has this probability distribution?

b. Make a histogram of this probability distribution table. Locate the fair price on the histogram. How does this compare to the mean of the distribution?

c. Complete the last column of this table, including the total.

Prize Value x	Probability $P(x)$	$x \cdot P(x)$
1	$\frac{1}{6}$	
2	$\frac{1}{6}$	
3	$\frac{1}{6}$	$\frac{3}{6}$
4	$\frac{1}{6}$	
5	$\frac{1}{6}$	
6	$\frac{1}{6}$	
Total	$\frac{6}{6}$	

d. Compare your total from Part c to the fair price from Part b. What do you notice?

2. The chart below gives the possible outcomes and their probabilities for a version of a scratch-off game played at McDonald's®. What is the fair price of one scratch-off card?

Prize	Probability
Win free fries worth 90¢	$\frac{1}{6}$
Win nothing	$\frac{5}{6}$
Total	$\frac{6}{6}$

3. Here is the table for another scratch-off game.

Prize Value	Probability
$1	$\frac{4}{10}$
$2	$\frac{2}{10}$
$3	$\frac{2}{10}$
$5	$\frac{1}{10}$
Win nothing	$\frac{1}{10}$
Total	$\frac{10}{10}$

a. What is the fair price of one scratch-off card?

b. Make a histogram of this probability distribution.

c. Estimate the balance point of the histogram. Compare this answer to the fair price of the card that you computed in Part a. What do you notice?

4. The mean of a probability distribution is called the **expected value** (*E.V.*).

   a. Explain why the expected value gives the fair price for a game if the probability distribution represents the chances of winning the various prizes in the game.

   b. Find the expected value of the probability distribution table for the sum of two dice. (See Activity 3 of Investigation 2 from Lesson 3, page 496).

5. By now you have a general procedure for finding the expected value of a probability distribution.

   a. Describe your method.

   b. Describe how to implement your method in an efficient way on your calculator or computer.

Checkpoint

Look back at your method for finding the expected value of a probability distribution.

ⓐ How is the method for finding the expected value of a probability distribution similar to the method for finding the mean of a frequency table? How is it different?

ⓑ What property of a probability distribution explains the difference in the methods?

Be prepared to share your group's analysis with the entire class.

▶ **On Your Own**

Find the expected value (fair price) of a ticket from the scratch-off game described in the following table.

Prize/Value	Probability
Free soft drink (89¢)	$\frac{15}{100}$
Free hamburger ($1.29)	$\frac{8}{100}$
T-shirt with restaurant logo ($7.50)	$\frac{3}{100}$
Movie passes ($15.00)	$\frac{1}{100}$
You lose! ($0.00)	$\frac{73}{100}$
Total	$\frac{100}{100}$

Modeling

1. A Las Vegas Keno ticket has the numbers from 1 to 80 on it. Twenty different numbers are drawn at random. Suppose a player chooses to mark one number on a Keno card. If that number is one of those drawn, the player wins $3.00 for a $1.00 bet.

   a. What is the probability the player wins $3.00?

   b. Is $1.00 a fair price for a ticket? Explain.

   c. What would be a fair price for a bet?

   d. A player also could choose to mark one number and bet $5.00. If that number is one of those drawn, the player wins $15.00. Is this wiser than betting $1.00?

2. Play the game below several times.

   In a game of matching, two people each flip a coin. If both coins match (both heads or both tails), Player B gets a point. If the coins don't match, Player A gets two points.

   a. What is the probability that Player A wins a round? That Player B wins?

   b. Explain why this is or is not a fair game.

3. The table below is copied from the back of a ticket in a scratch-off California lottery game.

Prize	Probability of Winning
$0.75	$\frac{1}{10}$
$2.00	$\frac{1}{14.71}$
$4.00	$\frac{1}{71}$
$10.00	$\frac{1}{50}$
$20.00	$\frac{1}{417}$
$250.00	$\frac{1}{1,000}$

   a. What is the probability of winning nothing with one ticket?

   b. What is the expected value of a ticket?

   c. The tickets in this lottery sell for $1.00 each. How much money does the California government expect to make if 1,000,000 tickets are sold?

4. A fast-food restaurant once had a scratch-off game in which a player picked just one of the four games on the card to play. In each game, the player stepped along a path, scratching off one of two adjacent boxes at each step. To win, the player had to get from start to finish without scratching off a "lose" box. Here's how the games on one card would have looked. (Of course, the words were covered until the player scratched off the covering.)

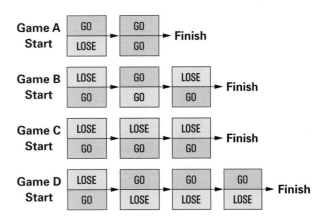

a. What is the probability of winning each game?

The prizes for the games were as follows.

    Game A: free food worth 55¢
    Game B: free food worth 69¢
    Game C: free food worth $1.44
    Game D: free food worth $1.99

b. What is the expected value of each game?

c. Which is the best game to pick if you just want to win something?

d. Which is the best game to pick if you want to have the largest expected value?

Organizing

1. Complete this sentence: In any probability distribution table,
$\sum P(x) =$ _____ because _____.

2. Using the indicated probability distribution tables from Investigation 2, page 514, find each of the following sums.
   ■ $\sum x$
   ■ $\sum P(x)$
   ■ $\sum x^2 \cdot P(x)$

   a. The table from Activity 2

   b. The table from Activity 3

3. Write an expression that uses a summation sign and gives the expected value of the probability distribution of a single roll of a die.

4. In a carnival game, players toss quarters onto a table marked with a grid. The length of a side of each square is 10 cm.

It's easy to be sure a quarter lands somewhere on the table, but exactly where is a fairly random event. If a tossed quarter lands entirely inside a square, the player wins a small prize. If the quarter touches a line, the player wins nothing. The game operator keeps the quarter in both cases. What should the prize be worth to make this a fair game? (**Hint:** Think in terms of where the center of the quarter must land so that the quarter does not touch a line.)

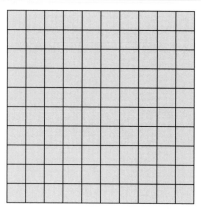

Reflecting

1. If your state has a lottery, investigate the amount of money bet, the amount of money paid in prizes, the operating costs of the lottery, the profit your state makes, and what your state does with the profits of its lottery. Compute the expected value of a ticket. Write a brief report summarizing your findings.

2. Why don't gambling games charge the fair price for playing? Why do people gamble when the price of playing a game is more than the expected value of the play?

3. According to the National Center for Health Statistics, in 1998 a newborn male in the United States could expect to live 73.8 years. A 20-year-old male could expect to live to the age of 75.0. A newborn female could expect to live to the age of 79.5, and a 20-year-old female to the age of 80.3. (Source: *World Almanac and Book of Facts 2001*. Mahwah, NJ: World Almanac, Inc. 2001.)

   a. What is meant, in this case, by the words "expect to live to the age of"?

   b. Why is the life expectancy for a 20-year-old greater than for a newborn?

   c. Do some research to find some of the reasons scientists give for why females can expect to live longer than males.

4. Investigate why young adult males have a higher death rate than young adult females. Do you believe it is fair to charge different rates for life insurance for men and for women? Is charging different rates legal in your state? What is the situation for automobile insurance?

Extending

1. This probability distribution table gives the probability of getting a given number of heads if a coin is tossed five times.

Number of Heads	Probability
0	$\frac{1}{32}$
1	$\frac{5}{32}$
2	$\frac{10}{32}$
3	$\frac{10}{32}$
4	$\frac{5}{32}$
5	$\frac{1}{32}$
Total	$\frac{32}{32}$

a. What is the expected number of heads if a person tosses five coins? Find the answer to this question in at least two different ways.

b. José says that the answer to Part a cannot involve half a head. How would you help him understand why it can?

2. Design a simple game that a baby-sitter could play with a young child. Make the game unfair. That is, one player has a better chance of winning than the other. Play your game with a friend and revise it if necessary. Play your game with a child. Did the child notice the game was unfair? In what ways did the child's understanding of probability differ from yours?

3. To help raise money for the local chapter of Big Brothers/Big Sisters, a college service organization decided to run a carnival. In addition to rides, the college students planned games of skill and games of chance. For one of the booths, a member suggested using a large spinner wheel with the numbers from 1 to 10 on it. The group considered three different games that could be played. The games are described in Parts a–c below. For each game, do the following:

- Make a table showing all possible outcomes and their probabilities.
- Calculate the expected value of the game, and compare that value to the price of playing the game.
- Determine if the game will raise money, lose money, or break even in the long run. If the game will not make money, recommend a change that the group might consider.

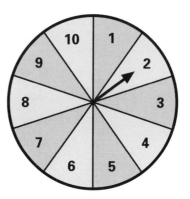

a. Double Dare To play this game, the player must pay the booth attendant $1. The player then chooses two numbers from one to ten and gives the wheel a spin. If the wheel stops on either of the two numbers, the attendant gives the player a prize worth $15.

b. Anything Goes For this game, the player can choose either one or two numbers. The player again must pay the attendant $1. After paying the attendant and choosing numbers, the player spins the wheel. If the wheel stops on a number the player chose, the attendant gives the player a prize. If the player selected only one number, the prize is worth $10. If the player selected two numbers, the prize is worth $5.

c. Triple Threat This game is a little more expensive to play. The player must pay $3 to spin the wheel. If the wheel stops on 1, 2, or 3, the player loses, receiving no prize. If the wheel stops on 4, 5, 6, or 7, the attendant gives the player a prize worth $2. If the wheel stops on 8, 9, or 10, the player gets a prize worth $6.

INVESTIGATION 3 Expected Value of a Waiting-Time Distribution

One of the waiting-time distributions you have constructed is for the waiting time to draw a brown "M&M's"® Chocolate Candy. The probability of success on a single trial is 0.3. The distribution table is shown below.

Drawing Candies

Number of Draws to Get First Brown One	Probability
1	0.3
2	$(0.7)(0.3)$
3	$(0.7)^2(0.3)$
4	$(0.7)^3(0.3)$
5	$(0.7)^4(0.3)$
6	$(0.7)^5(0.3)$
7	$(0.7)^6(0.3)$
8	$(0.7)^7(0.3)$
9	$(0.7)^8(0.3)$
10	$(0.7)^9(0.3)$
11	$(0.7)^{10}(0.3)$
12	$(0.7)^{11}(0.3)$
$\vdots$	$\vdots$

There should be an *infinite* number of rows in the table. It is possible, although definitely a rare event, that hundreds or thousands of candies could be drawn before the first brown one appeared.

1. In this activity and the next one, you will explore how to find the expected number of draws until a brown "M&M's"® Chocolate Candy appears. In other words, if a large number of people each draws candies until a brown one appears, what is the average number they draw?

   a. What happens if you try to use the $x \cdot P(x)$ procedure from Investigation 2 to find the expected value of this waiting-time distribution?

   b. Make an estimate of the expected value of the distribution by using just the first 25 rows of the table to compute the expected value. Keep at least 6 decimal places in all calculations.

   c. Is the real expected value larger or smaller than your estimate in Part b?

   d. How much would adding the 26th row change your expected value?

2. Place your estimated expected value from Activity 1, Part b in the appropriate space in a copy of the table below.

Expected Value of Waiting Times

Probability of a Success on Each Trial, p	Estimated Average Waiting Time (Expected Value)
0.10	10
0.20	
0.30	
0.40	
0.50	
0.60	
0.70	
0.80	
0.90	

a. Now the students in your class should regroup, if necessary, into seven groups, one for each of the remaining values of p in the table above. Each group should construct a waiting-time probability distribution table for its value of p. Tables should have at least 25 rows. Then estimate the expected waiting time for your value of p.

b. Get the expected values from each of the groups and fill in the rest of the table.

c. Using a copy of this coordinate grid, make a scatterplot of the data from your table.

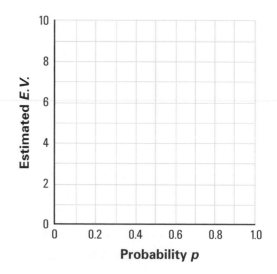

d. What is the expected value when $p = 1.00$? Plot the corresponding point on the scatterplot.

e. Describe the overall pattern relating probability p and expected value $E.V.$ Find an equation whose graph fits these points well.

f. According to your equation, how many candies would you expect to draw until you get a brown one?

3. Use your equation from Activity 2 to answer the following questions.

a. Suppose you want a sum of 6 on two dice. How many times do you expect to have to roll the dice?

b. About 25% of adults bite their fingernails. How many adults do you expect to have to choose at random until you find a fingernail biter?

c. There is about 1 chance in 14,000,000 of winning the California lottery with a single ticket. If you buy one ticket a week, how many weeks do you expect to pass until you win the lottery? How many years is this? If an average lifetime is 75 years, how many lifetimes is this?

Checkpoint

Suppose the probability of success on each trial of a waiting-time distribution is p.

ⓐ Write a formula for the expected value of this distribution.

ⓑ Explain what happens to the expected waiting time as p gets larger.

Be prepared to share your formula and reasoning with the class.

In Investigation 3, you discovered the formula for the expected value of a waiting-time distribution. It can be proved by twice using the method outlined in Extending Task 3 on page 529.

▶ On Your Own

Check your understanding of the expected value of a waiting-time distribution by completing the following tasks.

a. What information do you need about a waiting-time distribution in order to calculate the expected value of the distribution using the formula from the Checkpoint above?

b. Twenty percent of "M&M's"® Plain Chocolate Candies are red. How many candies would you expect to draw from a bag until you got a red one?

Modeling

1. Two statisticians have estimated that about 6% of all pennies go out of circulation each year. About 10,000,000,000 pennies are minted each year in the United States. Assume that was the number minted the year you were born.

   a. Complete this theoretical probability distribution table for the number of those pennies that go out of circulation each year. Add as many rows as you need to get to this year.

Circulation of Pennies from Your Birth Year

Years Since Your Birth	Number of Pennies That Go out of Circulation	Number of Pennies Still Left in Circulation
0	–	10,000,000,000
1	600,000,000	9,400,000,000
2		
3		
4		
5		
⋮		

   b. Write a *NOW-NEXT* equation describing the pattern of change in each of the last two columns.

   c. Approximately what percentage of the pennies minted the year you were born are still in circulation?

   d. Compute the average length of time a penny will stay in circulation using the formula from the Checkpoint on page 523.

   e. About how long will it (or did it) take for half of the pennies minted in the year you were born to go out of circulation? (This length of time is called the *half-life* of a penny.)

2. Painkillers often are given as shots to people who have sustained injuries. The time that it takes for a person's body to get the medicine out of his or her system varies from person to person. Suppose one person is given 400 mg of a medicine, and her body metabolizes the medicine so 35% is removed from her bloodstream each hour.

a. Complete the following chart.

Hours	Milligrams of Medicine Leaving the Blood	Milligrams of Medicine Left in the Blood
0	0	400
1		
2		
3		
4		
5		
6		

b. Is your table different from the other waiting-time distribution tables you have studied? Explain.

c. How long does the average milligram of medicine stay in the blood?

d. What is the approximate half-life of medicine in the blood? That is, how long does it take for half of the medicine to be gone?

3. In the 1986 nuclear reactor disaster at Chernobyl, in the former Soviet Union, radioactive atoms of strontium-90 were released. Strontium-90 decays at the rate of 2.5% a year.

a. What is the expected time it takes for a strontium-90 atom to decay?

b. Supposedly it will be safe again for people to live in the area after 100 years. What percentage of the strontium-90 released still will be present after 100 years?

4. Krypton-91 is a radioactive substance that gradually disintegrates. About 7% of krypton-91 disintegrates every second. Superman® is locked in a room with 5 grams of krypton-91. He can live only 30 seconds if he is near more than a gram of krypton-91.

   a. Will Superman make it?

   b. When will half of the krypton-91 be gone? What is the half-life of krypton-91?

Yes, comic fans, we know it's not the same Krypton.

Organizing

1. You have studied many different models for expressing relationships between quantitative variables: linear, exponential, power, quadratic, and trigonometric. When examining the pattern in your scatterplot from Activity 2 of Investigation 3 (page 522), which of these models could you immediately rule out as not reasonable? Explain your reasoning.

2. If the expected value of a waiting-time distribution is 6.5, what is the probability p of success on each trial?

3. James draws marbles one at a time from a bag of green and white marbles. He replaces each marble before drawing the next. If the probability that he gets his first green marble on the second draw is 0.24, what percentage of the marbles in the bag are green?

4. Waiting-time distributions can be modeled by exponential equations.

   a. Refer to Modeling Task 2, page 525. Write an equation that represents the number of milligrams of medicine left in the blood at any hour x.

   b. Refer to Modeling Task 1, page 524. Write an equation that represents the number of pennies minted in your birth year that are left in circulation x years later.

Reflecting

1. You now have several ways of finding the expected value of a waiting-time distribution. Describe each method. Will one of these methods be easier to remember than the others? Why?

2. Which of the following, if any, are correct interpretations of the expected value of a waiting-time distribution?

   ■ Half of the people wait longer than the *E.V.* and half shorter.

   ■ The *E.V.* is the most likely time to wait.

   ■ More than half of the people will wait longer than the *E.V.*

3. In Investigation 3, you discovered a surprisingly simple formula for the expected value of a waiting-time distribution. Write a summary of the methods that led to this discovery.

4. Most of the waiting-time distributions you investigated in this unit could be represented by formulas of the form $P(x) = pq^{x-1}$.

   a. For a given situation, what would the symbols, $P(x)$, p, and q represent? How are p and q related?

   b. What is the expected value of this distribution?

5. Ten billion pennies are minted in the United States every year.

   a. Make an estimate of how many rooms the size of your classroom are needed to hold 10,000,000,000 neatly stacked pennies.

   b. Count the number of pennies that can be found at your home. Make an estimate of the number of pennies in circulation in the United States. Don't forget the pennies in cash registers.

   c. What is the value of 10,000,000,000 pennies in dollars?

   d. If ten billion pennies were distributed equally among all of the people in the United States, how much money would you get?

   e. In what ways could a penny "go out of circulation"?

Extending

1. Examine this typical waiting-time probability distribution table.

Number of Trials x	Probability P(x)
1	$\frac{1}{6}$
2	$\left(\frac{5}{6}\right)\left(\frac{1}{6}\right)$
3	$\left(\frac{5}{6}\right)^2\left(\frac{1}{6}\right)$
4	$\left(\frac{5}{6}\right)^3\left(\frac{1}{6}\right)$
5	$\left(\frac{5}{6}\right)^4\left(\frac{1}{6}\right)$
⋮	⋮

   a. What waiting-time experiment could the above table be describing?

   b. Find the expected waiting time.

2. You have seen two ways to compute the expected value of the waiting-time distribution for rolling doubles. The first is to use the formula $\sum x \cdot P(x)$, which gives an **infinite series:**

$$1\left(\tfrac{1}{6}\right) + 2\left(\tfrac{5}{6}\right)\left(\tfrac{1}{6}\right) + 3\left(\tfrac{5}{6}\right)^{2}\left(\tfrac{1}{6}\right) + 4\left(\tfrac{5}{6}\right)^{3}\left(\tfrac{1}{6}\right) + 5\left(\tfrac{5}{6}\right)^{4}\left(\tfrac{1}{6}\right) + \cdots.$$

The second way is to use the formula you discovered in this investigation:

$$E.V. = \frac{1}{\tfrac{1}{6}} = 6$$

Since these two methods give the same expected value, you can set them equal. So,

$$6 = 1\left(\tfrac{1}{6}\right) + 2\left(\tfrac{5}{6}\right)\left(\tfrac{1}{6}\right) + 3\left(\tfrac{5}{6}\right)^{2}\left(\tfrac{1}{6}\right) + 4\left(\tfrac{5}{6}\right)^{3}\left(\tfrac{1}{6}\right) + 5\left(\tfrac{5}{6}\right)^{4}\left(\tfrac{1}{6}\right) + \cdots.$$

It's rather amazing that a series can keep going forever and still add up to 6.

a. Write the next three terms of the infinite series above.

b. Here is another example of an infinite series:

Since $\tfrac{1}{3} = 0.333333333\ldots$ (check by dividing 1 by 3), you can write

$$\tfrac{1}{3} = 0.3333333\ldots = \frac{3}{10} + \frac{3}{100} + \frac{3}{1,000} + \frac{3}{10,000} + \frac{3}{100,000} + \cdots.$$

Multiply both sides of $\tfrac{1}{3} = 0.3333333\ldots$ by 3. What do you conclude?

c. Write $\tfrac{2}{3}$ as an infinite series.

d. The following square is 1 unit on each side.

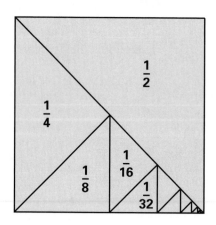

- What is the area of the square?
- Describe its area by adding the areas of the (infinite number of) individual triangles.
- What can you conclude?

3. Here is a way to find the sum of one kind of an infinite geometric series:

To find the sum of a series like $\frac{1}{2} + \frac{1}{4} + \frac{1}{8} + \frac{1}{16} + \frac{1}{32} + \cdots$, first set the sum of the series equal to S:

$$S = \frac{1}{2} + \frac{1}{4} + \frac{1}{8} + \frac{1}{16} + \frac{1}{32} + \cdots \qquad (*)$$

Multiply both sides by 2:

$$2S = 2\left(\frac{1}{2} + \frac{1}{4} + \frac{1}{8} + \frac{1}{16} + \frac{1}{32} + \cdots\right)$$

$$2S = \frac{2}{2} + \frac{2}{4} + \frac{2}{8} + \frac{2}{16} + \frac{2}{32} + \cdots$$

$$2S = 1 + \frac{1}{2} + \frac{1}{4} + \frac{1}{8} + \frac{1}{16} + \cdots \qquad (**)$$

Finally, subtract each side of equation (*) from the corresponding side of equation (**):

$$2S = 1 + \frac{1}{2} + \frac{1}{4} + \frac{1}{8} + \frac{1}{16} + \cdots$$
$$-S = \qquad \left(\frac{1}{2} + \frac{1}{4} + \frac{1}{8} + \frac{1}{16} + \frac{1}{32} + \cdots\right)$$
$$\overline{\qquad\qquad\qquad\qquad\qquad\qquad\qquad\qquad}$$
$$S = 1$$

In this example, both sides of equation (*) were multiplied by 2. With other infinite sums of this type, other numbers must be used.

a. Use the method above to find the sum of each of the infinite series below. Your first task will be to find the number to use for the multiplication.

■ $S = \frac{1}{3} + \frac{1}{9} + \frac{1}{27} + \frac{1}{81} + \cdots$

■ $S = \frac{7}{10} + \frac{7}{100} + \frac{7}{1,000} + \cdots$

b. Why doesn't the method work with the following infinite series?

$$\frac{1}{2} + \frac{1}{3} + \frac{1}{4} + \frac{1}{5} + \frac{1}{6} + \frac{1}{7} + \cdots$$

c. Show that the sum of the probabilities in the waiting-time distribution for rolling doubles is equal to 1.

Looking Back

In this unit, you have learned about probability distributions, specifically, the waiting-time (or geometric) distribution. You should be able to recognize a waiting-time situation and construct its probability distribution both theoretically and by simulation. You also should know how to find the average waiting time.

Among the concepts you investigated were probability distribution, graph of a probability distribution, expected value, fair price, Multiplication Rule, independent trials, and independent events. In this lesson, you will review and apply many of these ideas in new contexts.

1. In the "Simulation Models" unit in Course 1, one investigation focused on the population issues in China. In 2000, the population of China was more than 1,200,000,000. To control population growth, the government of China has attempted to limit parents to one child each. This decision has been unpopular in the areas of rural China where the culture is such that many parents desire a son.

   Suppose that a new policy has been suggested by which parents are allowed to continue having children until they have a boy. You might assume that half of all children born are boys, but the actual percentage is closer to 51%. For the following tasks, assume that the probability that a child born will be a boy is 0.51.

a. Describe a method using random digits to simulate the situation of parents having children until they get a boy.

b. Out of every 100 sets of parents, how many would you expect to get the first boy with the first baby? With the second baby? With the third baby?

c. Construct a theoretical probability distribution table for this situation, using a copy of the table below.

Number of Children to Get First Boy	Probability
1	
2	
3	
4	
5	
6	
7	
8 or more	

d. From your table, what is the average number of children two parents will have? From the formula for expected value, what is the expected number of children?

e. Explain whether the population will increase, decrease, or stay the same under this plan.

f. If this new policy were adhered to, what percentage of the population would be boys? Explain your reasoning.

2. Throughout the basketball season, Teri has maintained a 60% free-throw shooting average.

a. Suppose in the first game of the post-season tournament, Teri is in a one-and-one free throw situation. That is, if she makes a basket with her first shot, she gets a second attempt. Use an area model to determine:

- P(Teri scores 2 points)
- P(Teri scores 1 point)
- P(Teri scores 0 points)

b. What is the expected number of points Teri will score?

c. Explain how you could determine the probabilities in Part a without using an area model.

d. Now suppose that later in the game, Teri is in a two-shot foul situation. That is, she gets two attempts regardless of whether she makes the first shot. Determine the expected number of points Teri will score in this situation.

3. The Bonus Lotto game described below is similar to those played in many states. The jackpot starts at $4,000,000. On Saturday, 6 numbers from 1 through 47 are drawn. A seventh number, called the Bonus Ball, then is drawn from the remaining numbers. A player wins if the numbers he or she selects match the Bonus Ball and at least two of the numbers drawn.

The probabilities of winning various prizes are given in the following table:

Match	Winnings	Probability
6 of 6	$4,000,000	$\frac{1}{10,737,573}$
5 of 6 + bonus ball	$50,000	$\frac{1}{1,789,595}$
4 of 6 + bonus ball	$1,000	$\frac{1}{17,896}$
3 of 6 + bonus ball	$100	$\frac{1}{688}$
2 of 6 + bonus ball	$4	$\frac{1}{72}$
other	0	

a. What is the probability of winning nothing? Write your answer in decimal form.

b. What would be a fair price to pay for a ticket?

c. Bonus Lotto costs $2 to play. How much does the state expect to earn on every 1,000,000 tickets sold?

d. If the jackpot isn't won on the Saturday drawing, it grows by $4,000,000 for the next week. If you buy one ticket that second week, what is the probability of winning the jackpot?

e. What is a fair price for a ticket the second week?

f. What is the probability that a person who plays Bonus Lotto once this week and once next week will not win anything either week?

g. Suppose a person buys one ticket a week. What is the expected number of weeks he or she will have to wait before winning the jackpot? How many years is this?

h. The above table actually presents a simplified situation. In fact, if there is more than one winner, the $4,000,000 jackpot is shared. Explain why this fact makes the answer to Part b even smaller.

In this unit, you explored the mathematics behind waiting-time distributions. In the process you discovered the Mulitiplication Rule which can be used to find $P(A$ and $B)$ in certain situations.

a Write a general description of a waiting-time distribution. Include how to construct the probability distribution table, what the shape of the distribution looks like, and ways to find the average waiting time.

b For what kinds of problems and under what conditions should you use the Multiplication Rule to calculate probabilities?

Be prepared to compare your descriptions with those of other groups.

On Your Own

Write, in outline form, a summary of the important mathematical concepts and methods developed in this unit. Organize your summary so that it can be used as a quick reference in future units and courses.

Looking Back
at Course 2

Forests, the Environment, and Mathematics

In this course, you have continued your investigation of important mathematics and used it to analyze realistic situations. The mathematics you have studied includes matrices, systems of equations, coordinate models, transformations, correlation, linear regression, direct and inverse power models, quadratic models, network optimization, the geometry of mechanisms, trigonometric models, and waiting-time probability distributions. In this Capstone, you will pull together and

apply much of the mathematics you have learned, in order to analyze issues related to forests and the environment.

Forests are valuable for business and industry, recreation, and the maintenance of a healthy environment. Mathematics is used to help manage forests so that they can serve all these purposes most effectively.

Think About This Situation

A lumber company has submitted a proposal to begin logging operations in a nearby forest. While the owner of the property considers the proposal, the local community is debating its own concerns. The debate centers around three issues: economics, recreation, and the environment.

a What do you think is the major value of forests?

b Think of as many economic uses of forests as you can. Make a list.

c Make a list of recreational uses of forests.

d List as many environmental issues and benefits related to forests as you can.

INVESTIGATION 1 ► Forestry and Mathematics

Mathematics can be used in many different ways to help study and manage forests. Think about the mathematics you have studied in each of the units in this course (listed below). Brainstorm with your group about how the mathematics in each unit might be used in forestry. You may consider economic, recreational, or environmental aspects of forests. Your group should identify at least two ways to use the mathematics in each unit. Be prepared to share your group's thinking with the whole class.

1. *Matrix Models*
2. *Patterns of Location, Shape, and Size*
3. *Patterns of Association*
4. *Power Models*
5. *Network Optimization*
6. *Geometric Form and Its Function*
7. *Patterns in Chance*

Checkpoint

Different groups probably identified different ways in which the mathematical ideas in Units 1–7 might be used in forestry.

ⓐ For each unit, compare and discuss the ideas from different groups.

ⓑ Are there any big mathematical ideas or topics from this course that have not been applied to forestry? If so, is there any way they might be applied?

Be prepared to share your ideas with the entire class.

Your goal in this Capstone is to use mathematics to analyze certain aspects of forestry. At the end of the Capstone, you will prepare oral and written reports that will provide useful information for the community debate you considered in the "Think About This Situation" at the beginning of the Capstone. (Guidelines for the reports are given on page 554.)

As a group, take a quick look at Investigations 2 through 8 and then choose three to complete. (Investigation 7 counts as two.) Confirm your choices with your teacher before you begin. Investigations 2 through 7 include an optional "On Your Own" task. Individually, each group member should select and complete one of the "On Your Own" tasks from your group's investigations.

INVESTIGATION 2 Land Use Change in Rapid-Growth Areas

The United States Department of Agriculture (USDA) keeps track of how many acres of land are used for different purposes around the country. It publishes its data regularly so that planners and policy makers can make informed decisions. Examine the following USDA data. These data show how land use changed in 53 U.S. counties that grew rapidly in population during the period 1960 to 1970.

Transition Matrix of Land Use Change for 53 Rapid-Growth Counties, 1960–1970

Land Use in 1970

Land Use in 1960	1	2	3	4	5	6	7	8	9	10	11	12	Row Sum
1	6,315	106	363	2	48	149	4	45	14	55	20	6	7,127
2	108	1,514	100	0	18	12	0	7	3	5	2	1	1,770
3	128	70	1,387	2	180	145	3	23	16	58	32	13	2,057
4	2	0	2	185	0	1	0	1	0	2	0	0	193
5	26	14	74	2	6,309	143	2	34	6	33	9	4	6,656
6	0	0	1	0	0	1,449	0	2	0	2	0	0	1,454
7	0	0	0	0	0	17	49	2	1	12	0	0	81
8	0	0	0	0	0	0	0	780	1	0	0	0	781
9	0	0	0	0	0	0	0	0	113	0	0	0	113
10	0	0	8	0	0	0	0	0	0	391	0	0	399
11	0	0	7	0	0	1	0	0	0	0	645	0	653
12	5	1	14	0	1	2	0	2	0	0	3	435	463
Column Totals	6,584	1,705	1,956	191	6,556	1,919	58	896	154	558	711	459	21,747

Numbers in the matrix show land use change from 1960 (row) to 1970 (column), except for the diagonal, which shows use remaining unchanged. All numbers given are in 1,000 acres. The rows and columns are labeled with categories of land use according to the codes below.

Codes and Categories of Land Use

1. Cropland
2. Pasture & Range
3. Open/Idle
4. Farmstead
5. Forest
6. Residential
7. Urban Idle
8. Transportation
9. Recreation
10. Commercial, Industrial, Institutional
11. Water Bodies > 40 Acres
12. Miscellaneous

53 Rapid-Growth Counties

County	State	County	State	County	State
Madison	AL	Howard	MD	Morris	NJ
Santa Clara	CA	Montgomery	MD	Sussex	NJ
Santa Cruz	CA	Prince George's	MD	Cumberland	NC
Adams	CO	Plymouth	MA	Mecklenburg	NC
Arapahoe	CO	Macomb	MI	Wake	NC
Lee	FL	Washtenaw	MI	Portage	OH
Pasco	FL	Anoka	MN	Cleveland	OK
Sarasota	FL	Dakota	MN	Bucks	PA
Cobb	GA	Washington	MN	Chester	PA
De Kalb	GA	Jackson	MS	Collin	TX
DuPage	IL	Boone	MO	Dallas	TX
Lake	IL	Clay	MO	Denton	TX
Will	IL	Jefferson	MO	Harris	TX
Monroe	IN	St. Charles	MO	Tarrant	TX
Porter	IN	St. Louis	MO	Travis	TX
Johnson	KS	Sarpy	NE	Henrico	VA
Fayette	KY	Burlington	NJ	Waukesha	WI
Harford	MD	Monmouth	NJ		

Source: United States Dept. of Agriculture Economic Research Service, 1988.

1. Examine the transition matrix and the accompanying chart of rapid-growth counties from the USDA report.

   a. Are any of these rapid-growth counties near where you live?

   b. Discuss how to read the matrix. Give and explain two examples.

   c. Notice that the 5-3 (row 5, column 3) entry is not the same number of acres as the 3-5 entry. Explain why this is reasonable. What was the overall change in acreage between open land and forests?

   d. Why are there no negative entries in the matrix?

   e. What do the row and column sums mean in terms of land use? Explain.

   f. Which category of land use had the least change from 1960 to 1970? How can you tell?

   g. Which land use had the least *percentage* change from 1960 to 1970?

   h. Describe the change in land use related to forests from 1960 to 1970.

   i. Describe at least two other trends or patterns you see in this matrix.

2. The USDA transition matrix shows land use change in terms of number of acres. Another useful way to describe land use change is in terms of percent change.

Percent Change in Land Use

Land Use in 1970

	1	2	3	4	5	6	7	8	9	10	11	12
1	88.61	1.49	5.09	0.03	0.67	2.09	0.06	0.63	0.20	0.77	0.28	0.08
2	___	85.54	5.65	0	1.02	0.68	0	0.40	0.17	0.28	0.11	0.06
3	6.22	3.40	___	0.10	8.75	7.05	0.15	1.12	0.78	2.82	1.56	0.63
4	1.04	0	1.04	95.85	0	0.52	0	0.52	0	1.04	0	0
5	0.39	0.21	1.11	0.03	94.79	2.15	0.03	___	0.09	0.50	0.14	___
6	0	0	0.07	0	0	99.66	0	0.14	0	0.14	0	0
7	0	___	0	0	0	___	60.49	2.47	1.23	___	0	0
8	0	0	0	0	0	0	0	99.87	0.13	0	0	0
9	0	0	0	0	0	0	0	0	100.00	0	0	0
10	0	0	2.01	0	0	0	0	0	0	97.99	0	0
11	0	0	1.07	0	0	0.15	0	0	0	0	98.77	0
12	1.08	0.22	3.02	0	___	0.43	0	0.43	0	0	0.65	93.95

Land Use in 1960 (row axis label)

a. Examine the matrix above. Verify and explain the 1-5 entry of this matrix.

b. Now, by sharing the workload among members of your group, complete the remaining entries in the matrix.

c. Use the matrix and matrix multiplication to estimate the number of acres of land in each category in 1980, 1990, and 2010. Based on these estimates, describe at least two patterns in land use from 1960 to 2010.

d. Describe any limitations to the predictions you made in Part c. How could your predictions be improved?

3. Make a neat copy of your work on this investigation and file it at the location designated by your teacher. Examine the work filed by other groups in the class and compare their work to what you did. Write a question to at least one group asking its members to explain something about their work that you found interesting or that you did not understand. Answer any questions your group receives.

▶ **On Your Own**

Search your library or the Internet to find more recent USDA data on changes in land use. Use the more recent data to make a better estimate of land use for the year 2010.

INVESTIGATION 3 Valuing Urban Trees

Trees are an important part of urban landscapes. Whether in parks, in corporate plazas, or on city streets, trees are a valuable natural resource. In particular, they are worth money; and depending on size, location, and other factors, they can be worth a *lot* of money. Determining the dollar value of trees is one part of urban forestry.

1. Think of a tree somewhere near your school. Describe at least two ways that you think could be used to determine the dollar value of the tree.

2. The Council of Tree and Landscape Appraisers (CTLA) and the International Society of Arboriculture developed a method for valuing urban trees that has been used widely. This method uses the formula

$$V = B \times S \times L \times C$$

where V is the dollar value of the tree, B is the basic value based on cross-sectional area, S is the species value, L is the location value, and C is the condition value. Values for S, L, and C require the expert opinion of an urban forester, but the basic value B is computed using a table like the one on page 542. For this activity, only consider the basic value B.

   a. Use the table to find the basic value of a tree that has a 26-inch diameter.

   b. Let A represent the cross-sectional area and B the basic value of a tree. Write an equation showing the relationship between A and B. Describe the rate of change of the basic value as the cross-sectional area increases. Sketch a graph of the equation and describe how the rate of change is shown in the graph.

   c. Let D be the diameter of a tree with cross-sectional area A. Plot the (D, A) data and describe the relationship between D and A. Find and graph an equation that models the data.

   d. A key assumption has been made about the shape of a cross section, although the assumption is not explicitly stated in the table. Look at the data and think about trees. What assumption is being made about the shape of the cross section?

Basic Tree Value Determinations

Trunk Caliper or Diameter (in.)	Cross Section Area (in²)	Basic Value in Dollars (at $18/in.²)	Trunk Caliper or Diameter (in.)	Cross Section Area (in²)	Basic Value in Dollars (at $18/in²)
8	50.3	905	25	490.9	8,836
9	63.6	1,145	26	530.9	9,556
10	78.5	1,413	27	572.6	10,307
11	95.0	1,710	28	615.8	11,084
12	113.1	2,036	29	660.5	11,889
13	132.7	2,389	30	706.9	12,724
14	153.9	2,770	31	754.8	13,586
15	176.7	3,181	32	804.2	14,476
16	201.1	3,620	33	855.3	15,395
17	227.0	4,086	34	907.9	16,342
18	254.5	4,581	35	962.1	17,318
19	283.5	5,103	36	1,017.9	18,322
20	314.2	5,656	37	1,075.2	19,354
21	346.4	6,235	38	1,134.1	20,414
22	380.1	6,842	39	1,194.6	21,503
23	415.5	7,479	40	1,256.6	22,619
24	452.5	8,145			

Source: Council of Tree and Landscape Appraisers. *Guide for Establishing Values of Trees and Other Plants*, Revision 4. Savoy, Illinois: International Society of Arboriculture, 1979.

e. Let D be the diameter of a tree with cross-sectional area A. Using the assumption from Part d, write an equation showing the relationship between D and A. Compare this equation to your equation in Part c. Explain similarities and differences.

f. Use your equations from Parts b and e to predict the basic value of a tree that has a diameter of 80 inches. (An 80-inch-diameter tree is a very big tree, but, for example, it's less than half the diameter of the world's tallest tree. See Investigation 5.)

g. Combine your equation relating D and A from Part e with the equation relating A and B from Part b to get a single equation that shows the relationship between D and B. Use this equation to find the basic value of a tree with diameter 45 inches.

3. Make a neat copy of your work on this investigation and file it at the location designated by your teacher. Examine the work filed by other groups in the class and compare their work to what you did. Write a question to at least one group asking students to explain something about their work that you found interesting or that you did not understand. Answer any questions your group receives.

On Your Own

Locate a copy of the CTLA *Guide for Establishing Values of Trees and Other Plants* or the *Guide for Plant Appraisal*. Choose a tree near where you live. Use the valuation method described in the guide to estimate the value of the tree.

INVESTIGATION ▶ 4 Forests, the Greenhouse Effect, and Global Warming

The *greenhouse effect* refers to the warming effect of the Earth's atmosphere. Some gases in the atmosphere act much like the glass walls of a greenhouse, which let in sunlight from the outside but trap the resulting heat inside. In a similar manner, sunlight passes through the atmosphere and is absorbed by the Earth. The energy then radiates away as heat, but some of the heat is trapped by gases in the environment, thereby keeping the Earth's surface warm. It is the greenhouse effect that makes the Earth habitable. Without it, too much heat would escape into space and the Earth would be too cold to sustain life.

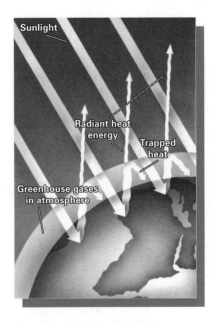

The main gases in the atmosphere responsible for the greenhouse effect, called "greenhouse gases," are water vapor, carbon dioxide, methane, nitrous oxide, ozone, and chlorofluorocarbons (CFCs).

1. Among all the greenhouse gases, water vapor and carbon dioxide have the highest levels of concentration in the atmosphere. Forests play a key role in determining the concentration level of carbon dioxide. Trees absorb vast quantities of carbon dioxide through the process of photosynthesis. Along with animals, trees also release carbon dioxide through respiration and decay. However, human activities, like fossil fuel consumption and destruction of forests, have begun to create extra carbon dioxide in the atmosphere. About one-fourth of the extra carbon dioxide resulting from human activity is attributed to deforestation.

   a. To examine the impact of deforestation, consider the case of Ecuador. In 1990, data showed that the annual rate of change in the area of forests in Ecuador was −1.99%. Assuming this rate of deforestation remains constant, write an equation using the words *NOW* and *NEXT* that shows how the area of forest land in Ecuador changes from year to year.

   b. In 1990, Ecuador had 155,760 square kilometers of forests. (Source: Food and Agriculture Organization of the United Nations. Forest Resources Assessment 1990–Global Synthesis, FAO Forestry Paper 124.) Use your equation from Part a to estimate the present area of forests in Ecuador.

c. Write an equation that uses the number of years elapsed since 1990 to estimate the area of forests in Ecuador. Use the equation to estimate the area of forests in Ecuador today and 7 years from now.

d. Locate current information giving the area of forests in Ecuador. Use that information to judge the accuracy of your equations from Parts a and c. If necessary, modify your models.

2. Recently, there has been concern that the greenhouse effect is being intensified artificially, resulting in so-called *global warming*. Is there a connection between increased carbon dioxide levels in the atmosphere and rising average temperatures over the whole planet? Is this a potentially serious environmental problem? Scientists have gathered data on carbon dioxide levels and average Earth temperatures over time in order to study these questions. Examine the table below, which shows some of these data. (Source: Intergovernmental Panel on Climate Change. *Climate Change 2001: The Scientific Basis.* Cambridge, U.K.: Cambridge University Press 2001.)

Global Warming?

Year	Carbon Dioxide Concentration (in parts per million ppm)	Temperature Deviation (in °C, compared to mean temperature from 1961–1990)
1960	317	−0.03
1965	320	−0.11
1970	326	−0.09
1975	331	−0.07
1980	338	0.05
1985	346	0.11
1990	354	0.20
1995	362	0.29
2000	371	0.34

a. Produce three scatterplots, one each for the (*year, carbon dioxide concentration*), (*year, temperature deviation*), and (*carbon dioxide concentration, temperature deviation*) data. Does there appear to be an association between any pair of variables?

b. Do you think any scatterplot reveals a linear association? Why or why not?

c. What statistical measures do you think would be helpful to compute? Explain your reasoning.

d. Write a brief report that you could submit to a group of concerned citizens. In the report, summarize your analysis of the relationship between carbon dioxide levels and changes in the Earth's climate. Include discussion of the *overall* trend during the period 1960–2000, trends *within* that period, use of the least squares regression lines for prediction, measures of correlation, and possibilities of a cause-and-effect relationship.

3. Make a neat copy of your work on this investigation and file it at the location designated by your teacher. Examine the work filed by other groups in the class and compare their work to that of your group. Write a question to at least one group asking its members to explain something about their work that you found interesting or that you did not understand. Answer any questions your group receives.

▶ **On Your Own**

In this investigation, you have examined just one of the greenhouse gases, namely, carbon dioxide. Choose one of the other greenhouse gases, and write a one-page essay describing it and its role in the greenhouse effect. Include in your essay its concentration in the atmosphere, how it is released into and absorbed from the atmosphere, and its particular environmental impact.

INVESTIGATION ▶ 5 Measuring Trees

Often, the first step in forest management is measuring the forest and the trees. For example, a lumber company planning to harvest a particular forest needs to know the height of the trees and the estimated board feet of lumber that the harvest will yield.

1. The tallest known tree in the world was discovered in 1963 by the National Geographic Society in Tall Trees Grove, Redwood National Park, California. The giant redwood tree, similar to the one pictured at the right, is 367.8 feet tall, has a circumference of 44 feet, and is 583 years old.

    a. How do you think the height of this tree was measured? List some possible measurement methods.

    b. Describe how trigonometry could be used to measure indirectly the height of the tree.

    c. Suppose that a forester wants to measure the height of a tree. She stands 50 feet from the base of the tree and uses a surveying instrument to determine that the angle of her line of sight to the top of the tree is 71.7°. Sketch a diagram of this situation and then find the height of the tree.

2. A *board foot* is a measure of volume of lumber. One board foot is a piece of lumber 1 foot long by 1 foot wide by 1 inch thick, or its equivalent. It is important to remember that, when measuring with board feet, you may not have pieces of lumber actually 1 ft × 1 ft × 1 in; instead you have the equivalent of such pieces. Lumber companies need to be able to estimate the number of board feet that a given log will yield. One of the most commonly used formulas for estimating board feet is the *Doyle Log Rule:*

$$B = \frac{L}{16}(D^2 - 8D + 16)$$

where B is board feet; D is the diameter, in inches, of the log inside the bark at the small end; and L is the length of the log in feet. For Parts a–e, consider only logs that are 16 feet long.

a. Rewrite Doyle's formula for the case of logs that are 16 feet long.

b. How many board feet will a 20-inch-diameter log yield (assuming the log is 16 feet long)?

c. Produce a graph and table for the quadratic equation you wrote in Part a. Describe the relationship between board feet and diameter shown by the graph and table. Is the rate of change of board feet with respect to diameter greater for large or small logs? How can you tell from the graph? From the table?

d. How many roots does the quadratic equation in Part a have? Find the roots and explain how you found them. What do the roots tell you about the kinds of logs for which Doyle's rule makes sense?

e. What information will you get if you solve the following quadratic equation?

$$620 = D^2 - 8D + 16$$

Solve the equation. Explain your solution method and the meaning of the answer.

f. So far you have used Doyle's rule for logs with a length of 16 feet and varying diameters. Now, rewrite Doyle's rule for logs that have a diameter of 28 inches and varying lengths. What kind of model do you get? Describe how board feet changes with respect to log length.

g. The equations you wrote in Parts a and f are quite different. They imply different rates of change for board feet as a function of diameter and for board feet as a function of length. Explain in your own words why it makes sense that board feet changes at a different rate with respect to diameter than it does with respect to length.

3. Make a neat copy of your work on this investigation and file it at the location designated by your teacher. Examine the work filed by other groups in the class and compare their work to yours. Write a question to at least one group asking its members to explain something about their work that you found interesting or that you did not understand. Answer any questions your group receives.

▶ On Your Own

Foresters in the field use a variety of handy and ingenious instruments to measure trees. For example, a *Biltmore stick* can be used to estimate tree diameter and a *relascope* can be used to estimate the sum of the cross-sectional areas of all trees in a particular region. Write a brief research report explaining how either the Biltmore stick or a relascope is used and why it works.

INVESTIGATION 6 Producing Wood Products

In the lumber business, there is a constant tension between harvesting and conserving trees. The profit of a lumber company depends not only on how many trees are harvested but also on the kind of wood products into which trees are converted.

1. Ketchikan Lumber Company in Alaska converts logs into particle board and into lumber such as two-by-fours and two-by-sixes. When the mill is running at peak capacity, it can turn out 400 units of wood products per week. The production cost for a unit of lumber is $30 and for a unit of particle board is $20. The owner wants to keep the mill running at full capacity while keeping production costs at $11,000 per week.

   a. Describe three different methods you could use to determine the number of units of each type of product that should be produced in a week.

   b. Choose one of your methods in Part a to find the number of units of lumber and of particle board that should be produced.

   c. Verify your solution in Part b using a different solution method.

   d. Of course, there are factors other than production costs that should be taken into account when setting production levels. Customer demand and profit per unit are two such factors. For example, suppose the weekly demand is for 100 units of particle board and 250 units of lumber. If the production levels stay as you calculated in Part b, what will happen to the company's inventory of these products? What would you suggest doing to correct this situation?

2. One type of saw used to produce lumber in sawmills is a *band saw*. A band saw blade is a long strip of metal, with teeth, that runs between two pulleys, similar to the side-view diagram at the right.

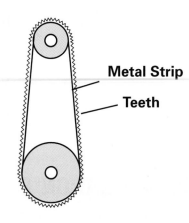

Metal Strip

Teeth

One band saw uses a blade that is 8 feet long in circumference. The larger pulley has a diameter of 12 inches and is attached to the saw motor, which has an angular velocity of 24,000 rpm. The smaller 6-inch pulley is adjusted to keep the blade tight.

a. At what angular velocity does the smaller pulley turn?

b. How many times will the blade revolve through its entire length in one minute?

c. The teeth should be sharpened after "traveling" 100,000 miles. About how often should the teeth be sharpened if the saw runs 6 hours per day?

3. Because of the potential danger in a sawmill, the machinery is tested thoroughly for reliability. Suppose the manufacturer of a band saw motor reports that the motor has about a 0.001 probability of failing each hour it is running.

a. Out of every lot of 10,000 motors produced, how many would you expect to fail in the first hour of operation?

b. Using the reliability estimate provided by the manufacturer, complete the following partial probability distribution table for the hour in which a sample motor fails.

Hour in Which Motor Fails	Probability
1	
2	
3	
500	
1,000	
1,500	

c. Sketch a graph of the probability distribution.

d. What is the expected number of hours until a motor fails?

e. The probability that a motor selected at random fails within the first 48 hours of operation is 0.046889. Find the probability that a motor fails within the first 49, 50, 51, and 52 hours. During what time frame would you consider failure of one of these motors to be a rare event?

f. For a given lot of 10,000 motors produced on the same assembly line, how many would you expect to fail within 52 hours of use? Do you think this is an acceptable number? Do you think 0.001 is an acceptable probability level for motor failure?

4. Make a neat copy of your work on this investigation and file it at the location designated by your teacher. Examine the work filed by other groups in the class and compare their work to what you did. Write a question to at least one group asking its members to explain something about their work that you found interesting or that you did not understand. Answer any questions your group receives.

▶ On Your Own

Referring to the situation in Activity 3, design a simulation model to estimate the probability that a motor fails within 50 hours. Extend the programming skills you developed in Unit 2 to develop a calculator or computer program to implement your simulation model. Run the program and compare your simulation results to the theoretical results you obtained in Part e of Activity 3. Explain any differences.

INVESTIGATION 7 Geographic Information Systems (GIS)

Many geographic features must be taken into account when studying a forest. *Geographic Information Systems* (GIS) compile all sorts of geographical data and use a variety of mathematical techniques to analyze the data. More and more foresters are using GIS in their analysis and management of forests. For example, Geographic Information Systems are used to locate fire towers, hiking paths, and microwave relay towers.

1. When considering where to locate fire observation towers, it is important to consider the location of key areas that are at-risk in a forest fire, such as cabin clusters, lodges, forest service buildings, and logging camps. In a Geographic Information System, these key areas and others are represented as points. Then the region containing the points is subdivided into a grid. One of the most useful ways to subdivide an area so that it can be described and studied systematically is to use a *Triangulated Irregular Network* (TIN). A TIN for a set of points is constructed by first drawing line segments between some of the points to form an outer boundary and then drawing additional segments to create a triangular network. This network is called a *triangulation*.

a. Below is a partial TIN for nine at-risk points in the Atika forest preserve. The outer boundary is completed, forming a polygon, and some of the additional segments needed to create the triangulation have been drawn. The TIN is not finished until all subdivided regions are triangles.

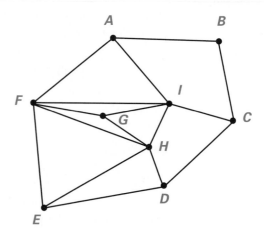

Working in pairs, complete the triangulation by adding more segments between points until the entire polygon is subdivided into triangles.

b. Compare your triangulation with that of other members of your group. Try to construct different triangulations. Is it possible to have two different triangulations for the same set of points?

c. There is a special triangulation, called a *Delaunay triangulation,* that is most useful. It is formed using triangles that are not "extreme." Extreme triangles are long and narrow, like triangles *FGH* or *FGI* in the triangulation from Part a. Brainstorm with your group to devise possible algorithms for producing a Delaunay triangulation from a partial TIN or a non-Delaunay TIN.

d. One way to modify any triangulation to get a Delaunay triangulation is as follows: Look for quadrilaterals formed by two triangles that share a side, where one of the triangles is "extreme." Swap diagonals in that quadrilateral if the result increases the size of the smallest of the six interior angles in the two triangles.

For example, in the triangulation you completed in Part a, consider the quadrilateral *FGHE*. Erase the existing diagonal, $\overline{FH}$, and replace it with $\overline{GE}$. Is the smallest angle of the two new triangles larger than the smallest angle of the original triangles? Does this diagonal swap eliminate the "extreme" triangle?

Continue swapping diagonals in this way until your triangulation in Part a has become a Delaunay triangulation.

2. Once a region has been triangulated, you can begin to analyze it. Continuing with the fire tower example, it is important to know which points are visible from different potential locations of a fire tower. Using known elevations of all vertices in the grid, you can construct a *visibility matrix* to help you decide where to place the fire tower. Consider the visibility matrix below, in which a "1" in a cell means that the two points are visible to each other.

Visibility Matrix

	A	B	C	D	E	F	G	H	I
A	1	1	0	0	0	1	1	0	1
B	1	1	1	0	0	0	0	0	1
C	0	1	1	1	0	0	1	1	1
D	0	0	1	1	1	0	0	1	0
E	0	0	0	1	1	1	1	1	0
F	1	0	0	0	1	1	1	1	0
G	1	0	1	0	1	1	1	1	1
H	0	0	1	1	1	1	1	1	1
I	1	1	1	0	0	0	1	1	1

a. Which points are visible from point *B*?

b. Assume that a fire tower will be located at one or more of the points *A* through *I*. If you can only build one fire tower, where would you put it? Why? Will one fire tower be enough to observe all nine points?

c. Compute the row sums and explain how they can help you answer the questions in Part b. (The row sums are called *visibility indices*.)

d. Where should fire towers be built so that all nine points can be observed and the fewest number of fire towers are built?

e. Describe the type of terrain that could yield a visibility matrix in which all the entries are 1s.

f. Construct the visibility matrix for a mountain range in the shape of a square-based pyramid, where the vertices in the matrix are the five vertices of the pyramid.

3. Geographic Information Systems also use vertex-edge graphs to represent and analyze geographic data.

a. The vertex-edge graph corresponding to a visibility matrix is called a *visibility graph*. Construct the visibility graph corresponding to the matrix at the beginning of Activity 2.

b. It is necessary to have line-of-sight communication between microwave transceivers for such things as telephone, television, and digital data networks. Referring to your visibility graph in Part a, suppose one transceiver is at vertex *A* and another is at vertex *D*. Since *A* and *D* do not have direct line-of-sight communication between them, relay towers will have to be built. Assume that relay towers will be built only at other vertices.

- What is the fewest number of relay towers necessary for line-of-sight communication between the transceivers at *D* and *A*? Where should the relay towers be built?

- Explain how finding a shortest path in the visibility graph between *D* and *A* provides an answer to the above questions.

c. Weighted graphs also are used to represent geographic data. The graph below represents hiking trails between lakes in Atika forest. The weights on the edges represent distances in miles.

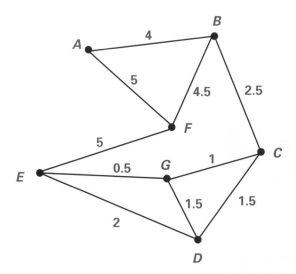

Because of weather and maintenance costs, it is not possible for the Forest Service to keep all trails open for the entire season. However, the Service does want to keep enough trails open so that it is possible to get from every lake to every other lake by some sequence of trails.

- What is the minimum number of miles of trails that the Forest Service must keep open? Sketch a network of open trails.

- Explain how you found your network of trails. Is this problem modeled with Euler paths, Hamiltonian paths, shortest paths, or minimal spanning trees?

4. In Activity 2, you were given a visibility matrix. To construct a visibility matrix, you need to figure out which points are visible from where. For example, consider the mountainous terrain represented by the diagram below.

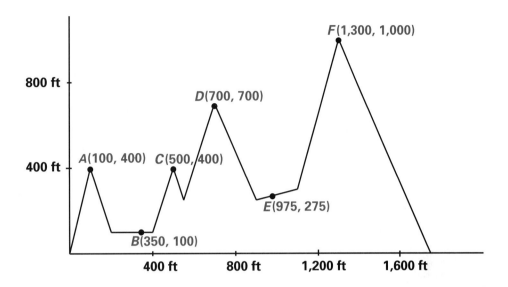

a. At which of the points *A–F* would you place a single fire observation tower so that you could see the most terrain? Explain.

b. Suppose there is an observation tower at point *F.* Can you see point *A* from the tower? Justify your answer. Give an argument supporting the claim that points *F, D,* and *A* lie on the same line of sight.

c. Can you see point *B* from point *D*?

d. Compute the slope of the line containing points *D* and *C.* Then compute the slope of the line containing points *C* and *B.* Explain how to use these slopes to answer the question in Part c.

e. Suppose that you work for the Forest Service and you are sent to point *D* with orders to keep a lookout for a renegade bear that has attacked several campers. In this extreme situation, your instructions are to shoot the bear with a tranquilizer dart if you get an open shot. You see the bear at a small lake located at point *E.* How far away is the bear? Do you think it is within range?

5. Make a neat copy of your work on this investigation and file it at the location designated by your teacher. Examine the work filed by other groups in the class and compare their work to what you did. Write a question to at least one group asking its members to explain something about their work that you found interesting or that you did not understand. Answer any questions your group receives.

> ## On Your Own

A Delaunay triangulation has the property that no vertices of the graph are enclosed by the circumscribing circle of any triangle. (There may be vertices *on* a circumscribing circle, however. In fact, the three vertices of a triangle *must* be on its circumscribing circle.) Verify this property by circumscribing circles for all the triangles in your triangulation from Part d of Activity 1. You may wish to use some geometry software to help you do this.

INVESTIGATION 8 Further Analysis

There are many other aspects of forests and the environment that you might investigate. Choose one of your ideas from Investigation 1 or from the "Think About This Situation" on page 536. Carry out a brief mathematical analysis of the idea. Specifically, you should formulate and answer at least two questions related to your idea.

REPORTS: Putting It All Together

Finish this Capstone by preparing two reports, one group oral report and one individual written report as described below.

1. Your group should prepare a brief oral report that meets the following guidelines:

   ■ Choose one of the investigations you have completed. Confirm your choice with your teacher before beginning to prepare your report.

   ■ Examine the work that other groups have filed on your investigation. Compare your work to theirs, discuss any differences with them, and modify your solutions, if you think you should.

   ■ Begin your presentation with a brief summary of your work in the investigation. Then explain your solutions to the various activities.

   ■ Be prepared to discuss alternative solutions, particularly those proposed by other groups that worked on your investigation.

   ■ Be prepared to answer any questions.

2. Individually, write a two-page report summarizing how the mathematics you have learned in this course can be used to understand issues related to forestry and the environment.

Checkpoint

In this course, you have continued to learn important mathematical concepts and methods and you have gained valuable experience in thinking mathematically. Look back over the investigations you completed in this Capstone and consider some of the mathematical thinking you have done. For each of the following habits of mind, describe, if possible, an example where you found the habit to be helpful.

a Search for patterns

b Formulate or find a mathematical model

c Collect, analyze, and interpret data

d Make and check conjectures

e Describe and use an algorithm

f Visualize

g Simulate a situation

h Predict

i Experiment

j Make connections—between mathematics and the real world and within mathematics itself

k Use a variety of representations—like tables, graphs, equations, words, and physical models

Be prepared to share your examples and thinking with the entire class.

Index of Mathematical Topics

Index of Contexts

Photo Credits

Cover, PhotoDisc; 319, NCSA, University of Illinois/Science Photo Library/Photo Researchers; 320, John Millar/Tony Stone Images; 322, Hank DeGeorge/*Chicago Tribune*; 325, Metro Group Editorial Service; 326, Karen Engstrom/*Chicago Tribune*; 327, Val Mazzenga/*Chicago Tribune*; 328, (bottom) FOTOPIC/West Stock; 329, Marjorie David/*Chicago Tribune*; 330, Alan Kehew/Western Michigan University; 331, John Lamb/Tony Stone Images; 333, *Chicago Tribune*; 335, Jim Steere/courtesy Chicago Symphony Orchestra; 338, courtesy Joseph Kruskal from *Discrete Mathematics* by Susan Epp, Clarendon Press, Oxford, 1976; 340, Robert E. Daemmrich/Tony Stone Images; 342, (top) GAMES OF THE WORLD by Frederic B. Grunfeld, © 1975 by Product Development International Holding n.v., reprinted by permission of Henry Holt and Company, Inc., (bottom) Ron Bailey/*Chicago Tribune*; 344, Aaron Haupt; 345, Stephen Mason/West Stock; 346, *Graph Theory 1736-1936*, Norman L. Biggs, E. Keith Lloyd, Robin J. Wilson, Clarendon Press, Oxford, 1976 "by permission of Oxford University Press"; 348, Frank Hanes/*Chicago Tribune*; 352, courtesy Apple Computer; 353, Charles Thatcher/Tony Stone Images; 356, John Kuntz/*Chicago Tribune*; 359, Wadsworth Publishing Co., Brooks/Cole, University of Texas; 360, Aaron Haupt; 362, Martin Griff; 364, Melanie Carr/West Stock; 367, Dan Feicht; 368, (top left) AP/Wide World Photos, (top right) Roy Hall/*Chicago Tribune*, (bottom right) Bob Thomason/Tony Stone Images; MGA: Tom McCarthy; 370, Mulcahy/Wieting/*Chicago Tribune*; 371, Bobbi Lane/Tony Stone Images; 372, AFP/CORBIS; 373, *Geometry, Second Edition*, Scott Foresman & Co., 1987; 377, (top) Jack Demuth, (bottom left) Mark Saperstein, courtesy Emily Sacca, (bottom right) Jim Brown; 378, Jack Demuth; 379, Jim Brown; 383, UPI/Bettmann/CORBIS; 387, Jack Demuth; 388, Fran Brown; 389, (top) UPI/J. Dickerson, (bottom) Tony Freeman/PhotoEdit; 390, M. Ferguson/PhotoEdit; 391, Don Casper/*Chicago Tribune*; 396, Tony Freeman/PhotoEdit; 398, Texas Instruments Inc, Dallas, Texas; 400, Walter Neal/*Chicago Tribune*; 401, Jack Demuth; 402, Carl Hugare/*Chicago Tribune*; 406, (right) Peter Timmermans/Tony Stone Images, (left) Reuters/Bettmann/CORBIS; 407, NASA; 408, UPI/Bettmann/CORBIS; 409, AP/Wide World Photos; 411, Pat O'Hara/Tony Stone Images; 413, Jack Demuth; 415, John Laptad/West Stock; 419, Sears Roebuck and Co, Chicago, IL; 424, AP/Wide World Photos; 425, Troy-Bilt courtesy Garden Way; 428, Jim Brown; 430, Linn Sonder Electric; 432, Ken Anderson/West Stock; 435, Chicago Co. Fair; 436, Dave Nystrom/*Chicago Tribune*; 438, Aaron Haupt; 443, AP/Wide World Photos; 449, Charles D. Winters/Photo Researchers; 451, CORBIS; 452, (top) Techsonic Industries, (bottom) Telegraph/FPG; 455, courtesy Didax Educational Resources; 456, 457, Hasbro, Inc., all Rights Reserved, used with permission, photos by Jack Demuth; 462, ETHS Yearbook Staff; 465, Ed Wagner/*Chicago Tribune*; 467, Jack Demuth; 469, Audi of America, Inc.; 471, ETHS Yearbook Staff; 474, *Chicago Tribune*; 475, ETHS Yearbook Staff; 477, Audi of America, Inc.; 478, (left) Reuters NewMedia, Inc./CORBIS, (right) Anna Belluomini; 480, ETHS Yearbook Staff; 481, Gillette Personal Care Products; 483, courtesy Robert and Mary Cooper; 484, Mark Saperstein; 485, Jack Demuth; 489, David Young-Wolff/PhotoEdit; 490, (top) United Press Photo, (bottom) Tim Defrisco/Allsport; 492, Jack Demuth; 495, United Press Photo; 498, Jim Brown; 503, Mario Petitti/*Chicago Tribune*; 506, Ed Pritchard/Tony Stone Images; 508, Walter Kale/*Chicago Tribune*; 511, (top) © 1997 Nintendo, all rights reserved, (bottom) Vic Bider/PhotoEdit; 519, Michael Newman/PhotoEdit; 520, Carl Hugare/*Chicago Tribune*; 523, ETHS Yearbook Staff; 524, 525, AP/Wide Word Photos; 526, DC Comics © 1997, all rights reserved, used with permission; 527, UPI; 530, 531, AP/Wide World Photos; 535, Tom Bean/Tony Stone Images; 536, Stephen Studd/Tony Stone Images; 538, Terry Donnelly/Tony Stone Images; 543, Carlyn Iverson; 545, Tom & Pat Leeson/West Stock; 547, Lon E. Lauber M.R./West Stock; 549, Earl Roberge/West Stock; 551, Tony Stone Images.